STEM to STREAMS

STEM to STREAMS

Toward a More Equitable Vision of
STEM Education

Edited by
Douglas Huffman
Kelli R. Feldman
Imogen R. Herrick

ROWMAN & LITTLEFIELD
Lanham • Boulder • New York • London

Rowman & Littlefield
Bloomsbury Publishing Inc, 1385 Broadway, New York, NY 10018, USA
Bloomsbury Publishing Plc, 50 Bedford Square, London, WC1B 3DP, UK
Bloomsbury Publishing Ireland, 29 Earlsfort Terrace, Dublin 2, D02 AY28, Ireland
www.rowman.com

Copyright © 2025 by Douglas Huffman, Kelli R. Feldman, and Imogen R. Herrick

All rights reserved. No part of this publication may be: i) reproduced or transmitted in any form, electronic or mechanical, including photocopying, recording or by means of any information storage or retrieval system without prior permission in writing from the publishers; or ii) used or reproduced in any way for the training, development or operation of artificial intelligence (AI) technologies, including generative AI technologies. The rights holders expressly reserve this publication from the text and data mining exception as per Article 4(3) of the Digital Single Market Directive (EU) 2019/790.

British Library Cataloguing in Publication Information Available

Library of Congress Cataloging-in-Publication Data Available

ISBN 978-1-4758-7423-5 (cloth) | ISBN 978-1-4758-7425-9 (electronic)

For product safety related questions contact productsafety@bloomsbury.com.

∞™ The paper used in this publication meets the minimum requirements of American National Standard for Information Sciences—Permanence of Paper for Printed Library Materials, ANSI/NISO Z39.48-1992.

Contents

Preface   vii
Douglas Huffman, Kelli R. Feldman, and Imogen R. Herrick

1   Estuaries of Learning: Cultivating STREAMS Education
through Transdisciplinary Teaching and Learning   1
Imogen R. Herrick and Michael Lawson

2   Digital Humanities and STREAMS: Pre-service Teachers'
Exploration of Innovative Curricular Pathways   23
Britta Bletscher and Heidi L. Hallman

3   Computer Science Curriculum for Culturally and
Linguistically Diverse Students   37
*Clare Baek, Sharin Jacob, Dana Saito-Stehberger,
Leiny Yesenia Garcia, Santiago Ojeda-Ramirez, and Mark Warschauer*

4   Child-Robot Musical Theater for Diverse and Inclusive
STREAMS Education for Young Children   51
Koeun Choi and Myounghoon Jeon

5   Critical Service-Learning: A Transformational Framework for
Integrating STEM and Social Justice   73
Cara Eleonora Daza

6   Developing a Conceptual Framework for Early Childhood
Streams Instruction: Integrating Social Emotional Competencies   89
Amanda Bennett

7 Building a Diverse Stream of Engineers through Teacher
   Research Experiences in Engineering: The IDEA-BioE Project   105
   Meagan M. Patterson, Prajnaparamita Dhar, and Douglas Huffman

8 Beyond the Technocratic View of Engineering   121
   Gillian Roehrig

Afterword: Confluence   137
Douglas Huffman, Imogen R. Herrick, and Kelli R. Feldman

About the Editors and Contributors   141

Preface

Douglas Huffman, Kelli R. Feldman, and Imogen R. Herrick

The whole is greater than the sum of the parts.

Aristotle is often attributed with the phrase "The whole is greater than the sum of the parts," although what he wrote was that "The whole is something besides the parts" (Aristotle, 980a). The idea that the whole is greater than the parts has merit today as we think about STEM fields and how more and more STEM disciplines are merging. The integration and interrelationships among disciplines can create new fields, such as nanoscience and technology, that are greater than the sum of the separate disciplines. These new fields are often referred to as *transdisciplinary* (Huffman et al., 2020).

In the classroom, transdisciplinary learning describes experiences where concepts and skills from all disciplines are applied during authentic problem-based learning (English, 2016). Traditionally, transdisciplinary learning is discussed in terms of STEM education (e.g., Akerson et al., 2018), where the disciplines of science, technology, engineering, and mathematics are collectively drawn upon for problem-solving. However, in 2007, a group at the Americas for the Arts National Policy Roundtable grappled with the need to attract and recruit students to study STEM education. The group focused on how integrating arts could open opportunities for STEM education to be more welcoming to students, particularly those not typically represented in STEM. The group discussed adding the arts to STEM education (Perignat & Katz-Buonincontro, 2019). Simultaneously, arts was added to STEM by the Rhode Island School of Design, and they began using the term STEAM to highlight the role of art and design in transdisciplinary K-20 education.

Over the next several years, the concept of STEAM began to take hold; however, the benefits of adding the arts to STEM education were deliberated. It was argued that integrating arts across the STEM disciplines would make STEM education more accessible and would contribute to growing a more diverse and innovative workforce (National Academies of Science, Engineering, and Medicine [NASEM], 2016). However, these aims have yet to materialize. In recent years, the addition of the arts to STEM has created new cultures and practices for STEM teaching and learning that go far beyond technocratic goals—for example, hip-hop and reality pedagogies (Emdin et al., 2016). It is through these new cultures and practices around STEM learning that new ways of thinking about what it means to do STEM emerge, and idea of evolutions are needed for more individuals to find meaning in doing STEM. We see the idea of STREAMS as a needed evolution of STEAM, where it is not implicit but explicit that political, social, historical, and cultural ways of knowing, including and beyond art, are equally important lenses to make sense of the world around us.

But what will the future of STEM, STEAM look like, or how can STREAMS be achieved? In this book, we feature chapters from scholars in STEM education, educational leadership, and educational research. The chapters explore how their fields of education integrate with STEM to produce new disciplines that have not traditionally been part of STEM. The book is organized around the idea of "STREAMS" to conceptualize the future of STEM and STEAM education. STREAMS includes (1) Science, (2) Technology, (3) Reading/ELA, (4) Engineering, (5) Arts, (6) Mathematics, and (7) Social Sciences.

This book seeks to push the STEM fields forward using the metaphor of STREAMS. At times, the STREAMS are more like gently babbling brooks that meander along in a relaxing manner. At other times, the STREAMS are overflowing and gushing with energy. Each of these STREAMS can help to create new *transdisciplinary* concepts that are greater than the STEM field alone. The STREAMS metaphor is hopefully a useful way to think about the myriad ways one could combine, integrate, or merge with the different STEM fields.

An important reason to broaden the field from STEM to STREAMS is that the integration of STEM with other threads of disciplines can help to broaden the engagement of more students in STEM, particularly students from backgrounds that are underrepresented in STEM fields (National Science Board, National Science Foundation, 2024). Providing transformative learning opportunities in STREAMS can support students in "reauthoring their rights" to STEM (Calabrese Barton & Tan, 2020). The concept of "reauthoring rights" to STEM pushes back on the idea that women and underrepresented students should "fit" into traditional STEM fields, rather than work to change the status quo. The field of STEM education has focused most of its effort on trying to help underrepresented students fit into STEM.

STEM education has traditionally focused on developing skills for the STEM workforce pipeline, often overlooking students who might pursue STEM fields through different paths (Aikenhead, 2018). STREAMS education offers a dynamic focus, where learning experiences activate students to tackle local challenges and imagine just and sustainable futures. Unlike STEM, the incorporation of disciplines in the humanities within the idea of STREAMS engages a humanistic approach to transdisciplinary teaching and learning, one that intends to empower students to connect *during* learning through their experiences and histories. These connections during learning are particularly powerful for students with backgrounds that have been historically minoritized (e.g., Kokka, 2019; Moll et al., 1992; Morales-Doyle, 2024). Additionally, positioning students as capable and dynamic solution creators supports them in reauthoring their rights and responsibilities as members of a community who can shape the future (e.g., Herrick et al., 2022; Morales-Doyle, 2017). These dimensions of STREAMS acknowledge the different types of learning experiences needed for students to tackle local challenges and create space to cultivate new cultures around what counts as learning and what purpose learning holds and for whom.

As we look across the chapters in this book, we observe the clear and sustained presence of technology across the experiences described. As such, we begin by exploring how teachers and students use technology to reveal aspects of community, relationships, or ideas for action.

Chapter 1: Herrick and Lawson provide a foundation for considering the role of technology in STEM. Estuaries are used as metaphors to envision how various streams branch off from each other and serve as an undercurrent to connect STEM. The STEM field tends to think of technology as one of the big four ideas, but Herrick and Lawson push our thinking on what technology means and how it is the connective tissue for transdisciplinary learning. They describe two flexible, justice-centered routines used across K–12 classrooms.

Chapter 2: Bletscher and Hallman explore STREAMS through the intersection of technology and humanities in their chapter by describing a new field called digital humanities. Using tools such as KnightLab and Voyant, they demonstrate how digital humanities enrich traditional ELA curriculum by fostering deeper, interdisciplinary engagement and preparing all students for a digitally interconnected world.

Chapter 3: Baek and colleagues emphasize the importance of embedding reading, writing, and disciplinary literacy goals into STEM curricula. They explore how computational thinking, when combined with literacy, can enhance students' ability to communicate and collaborate on solving real-world issues. Using the programming platform Scratch, this chapter examines a culturally relevant computer science curriculum designed to foster creativity and computational thinking while also developing students' literacy skills.

Chapter 4: Choi and Jeon use social robots as another means of engaging students in STEM. The growing prevalence and diversity of social robots offer opportunities to promote inclusive STREAMS education. Theater activities co-created by children and social robots provide a meaningful learning context to explore the world. The project aims to incorporate children's interests and experiences by integrating artistic formats, including drama, dancing, music, and drawing, and interactive storytelling with social robots.

Chapter 5: Daza on Critical Service Learning explores a unique STEM service-learning project in Colombia, South America. Critical service-learning is a pedagogical approach that prioritizes relational, equitable, and action-oriented principles, integrating STEM and social justice to foster transformative educational experiences. The chapter then examines the efficacy of critical service-learning as a tool for promoting equity within STEM education.

Chapter 6: Bennett's chapter proposes a new model of early childhood education and STEM. The model draws on STEM, STEAM, and social emotional competencies (SEC) to provide a developmentally appropriate approach that integrates character skills development and academic tools more intentionally to produce lifelong learners and productive members of society. The integration of STREAMS and SEC represents a natural alignment with ECE's fundamentals, in that students who engage in STREAMS education can become capable of successfully managing emotional processes, employing social/interpersonal skills, and utilizing appropriate cognitive skills, including critical thinking, and creativity.

Chapter 7: Patterson, Dhar, and Huffman describe a BioEngineering research project for teachers as a means to increase the access and participation of women in STEM. Women are in high demand in the field of engineering but often do not feel welcomed in engineering settings. STREAMS is used as a way to consider building new pipelines or "streams" of engineers entering the field. In-service and pre-service teachers alike worked in Bioengineering labs over the summer to learn about cutting-edge bioengineering concepts and to work in engineering labs to create welcoming engineering spaces. The laboratory experience is then followed by teachers creating new modules that can be used in middle school science.

Chapter 8: Finally, Professor Gillian Roehrig provides a chapter on concerns about the use of STREAMS for STEM education. The chapter questions whether or not attempting to expand the view of STEM has gone too far. The field has added "the arts" to STEM and created the idea of STEAM. Many schools have embraced STEAM, but this new concept of STEAM should be used cautiously. This book pushes even further and provides a warning for how one might push the integration even further. One way to envision STREAMS is not to attempt to simultaneously include each area at the same time—S.T.R.E.A.M.S., but rather to think of each branch of the stream as another way to diversify the experience in engineering.

Overall, we hope this book will inspire new ways to think about STEM education and to consider how the STREAMS metaphor can lead to the creation of new rivers, creeks, brooks, or cascades. These new branches of the stream can help reach *all* students—both underrepresented and minoritized students. Each stream has unique characteristics. Each stream meanders in its own way, just as each learner enters the field of STEM in their own unique way. By considering various streams and rivers, we can begin to envision a watershed where the rivers and all the tributaries help to provide opportunities for all.

More importantly, each branch of the stream can welcome students into STEM in ways we can never predict. We should help students take back STEM and "reauthor their rights" to STEM education (Calabrese Barton & Tan, 2020). Students can reauthor their rights to create a unique environment that fits their unique needs. Underrepresented and minoritized youth do not all fit into STEM. Why should all students fit into the traditional STEM fields? It is time to create new streams of knowledge, new pathways to enter STEM, and new ways to recreate environments that are more diverse, equitable, and inclusive, where all students truly feel like they belong.

REFERENCES

Aristotle. (1933, 1989). *Aristotle in 23 Volumes, Vols. 17, 18* (H. Tredennick, Trans.). Harvard University Press; William Heinemann Ltd.

Calabrese Barton, A., & Tan, E. (2020). Beyond equity as inclusion: A Framework Of "Rightful Presence" For Guiding Justice-oriented Studies In Teaching And Learning. *Educational Researcher, 49*(6), 433–440.

Emdin, C., Adjapong, E., & Levy, I. P. (2021). On science genius and cultural agnosia: Reality pedagogy and/as hip-hop rooted cultural teaching in STEM education. *The Educational Forum, 85*(4), 391–405.

English, L. D. (2016). STEM education K–12: Perspectives on integration. *International Journal of STEM Education, 3*(1), 3.

Herrick, I. R., Lawson, M., & Matewos, A. M. (2022). Through the eyes of a child: Exploring and engaging elementary students' climate conceptions through photovoice. *Educational and Developmental Psychologist, 39*(1), 100–115.

Huffman, D., Ristvey, J., Morrow, C., & Deal, M. (2020). Integrating nanoscience in high school science: Curriculum models and instructional approaches. In K. Sattler (Ed.), *21st century nanoscience: A handbook*. Taylor & Francis.

Kokka, K. (2019). Healing-informed social justice mathematics: Promoting students' sociopolitical consciousness and well-being in mathematics class. *Urban Education, 54*(9), 1179–1209.

Moll, L., Amanti, C., Neff, D., & Gonzalez, N. (1992). Funds of knowledge for teaching: Using a qualitative approach to connect homes and classrooms. *Theory into Practice, 31*(2), 132–141.

Morales-Doyle, D. (2017). Students as curriculum critics: Standpoints with respect to relevance, goals, and science. *Journal of Research in Science Teaching, 55*(5), 749–773.

Morales-Doyle, D. (2024). *Transformative science teaching: A* catalyst *for* justice *and* sustainability. Harvard Education Press.

National Academies of Sciences, Engineering, and Medicine. 2016.*Barriers and Opportunities for 2-Year and 4-Year STEM Degrees: Systemic Change to Support Students' Diverse Pathways*. Washington, DC: The National Academies Press. https://doi.org/10.17226/21739.

National Science Board, National Science Foundation. (2024). *Science and engineering indicators 2024: The state of U.S. science and engineering*. NSB-2024-3. https://ncses.nsf.gov/pubs/nsb20243

Perignat, E., & Katz-Buonincontro, J. (2019). STEAM in practice and research: An integrative literature review. *Thinking Skills and Creativity, 31,* 31–43.

Chapter 1

Estuaries of Learning

Cultivating STREAMS Education through Transdisciplinary Teaching and Learning

Imogen R. Herrick and Michael Lawson

Emdin (2021) explains, "STEM is not a collection of academic subjects. It is not a field of study or an approach to curriculum. It is an idea" (p. 3). In this sense, STREAMS is also an *idea* with endless possibilities and situational constraints. In this chapter, we delve into the concept of STREAMS, with a focus on technology, to think about how teachers and students can engage in knowledge-building around complex local phenomena, employ imaginative and critical thinking across disciplines, and discover opportunities to act on their new understandings. We begin exploring the *idea* of STREAMS by first examining it through the phenomena of natural streams and the ecosystems they create through movement and merger. This practice, called biomimicry, looks to the wisdom of nature to draw inspiration for potential structures and functions of STREAMS as an *idea* for education. We then explore how nature's wisdom shapes the *idea* of STREAMS and technology's pivotal role in affording such learning opportunities. Finally, we offer two *justice-centered flexible routines*, Photovoice and Community Science Data Talks, to demonstrate how to integrate practical technologies in ways that can engage and sustain the *idea* of STREAMS through small-scale shifts in the classroom.

STREAMS CLASSROOMS AS ESTUARIES OF LEARNING

Just as the acronym STREAMS is an idea, it is also a word that brings to mind the idea of streams flowing in nature. This duality invites exploring and innovating through *biomimicry*, where we look to natural streams for inspiration in understanding how STREAMS education can be actualized and sustained in classrooms.

Across landscapes, natural streams flow and merge with each other and, eventually, with larger bodies of water. These movements and mergers provide guidance on what are needed when working toward employing a STREAMS approach to problem-solving in classrooms (e.g., English, 2016; Kelley & Knowles, 2018). For instance, the *flexibility* of water moving and merging in streams mimics the *flexibility* required to move with student ideas and support mergers of content knowledge during learning. Observing the movements and mergers of streams affords opportunities to apply these observations to the movements and mergers needed to achieve transdisciplinary learning.

To begin, *individual freshwater streams* represent the distinct disciplines within K12 curricula—such as science, technology, reading/ELA, engineering, arts, mathematics, and social sciences—each carrying its own unique concepts and skills. As these streams move across the landscape, they encounter opportunities to *intersect* and *merge*, similar to how various academic disciplines can converge and integrate during the learning process. In nature, the place where freshwater streams intersect and merge into larger bodies of water is called a *confluence*. At these places of intersection and merger, there is an increase in the volume and intensity of water flow, which can alter the stream's direction, behavior, and ecological characteristics.

As a natural phenomenon, *confluences* can serve as a metaphor for understanding the levels of integration of STEM disciplines in educational settings and the potential outcomes of such integrations. For instance, the intersecting of streams can mimic a multidisciplinary approach, where disciplines are taught separately but around a common theme. While the merging of multiple streams mimics an interdisciplinary approach, where disciplines are deeply interconnected, significantly enhancing and enriching the understanding of specific knowledge and skills (e.g., English, 2016). Observing the movements and mergers provides insights; however, these observations lack the ability to develop the nuance needed to capture the full complexity around creating STREAMS learning for a diversity of students. This is because confluences involve merging bodies of freshwater and are not sites that can support the full diversity of aquatic life.

Natural streams also intersect and merge with the ocean's saltwater in *estuaries,* which provide a richer context for biomimicry's ability to inspire STREAMS education. In an estuary, the blending of freshwater from streams with saltwater from the ocean creates fluctuating levels of salinity that can propagate life and new possibilities for life. As such, estuaries are some of the most vibrant ecosystems on Earth. *We imagine STREAMS classrooms as estuaries, where flexibility, movement, and merger coalesce to cultivate transdisciplinary learning opportunities from which a diverse array of students and teachers can flourish and thrive* (seefigure 1.1).

In a STREAMS learning opportunity, we imagine the *ocean's saltwater* as complex, real-world, justice-centered issues in students' lives that are not traditionally engaged during classroom learning (e.g., Borrero & Sanchez, 2017; Morales-Doyle, 2017). Through the ocean's tides, over time, salt and freshwater mix in *estuaries* through small interactions that propagate and expand in scale. Similarly, over time, locally relevant issues provide teachers and students with flexible and interconnected opportunities to develop knowledge and build relationships around phenomena impacting their lives and each other. Blending STREAMS disciplinary learning with real-world issues, inspired by the natural wisdom of *estuaries*, is a justice-centered (e.g., Morales-Doyle, 2017) educational approach that empowers students to address complex global challenges with imaginative and critical thinking and local problem-solving skills.

TRANSDISCIPLINARY TEACHING AND LEARNING: FROM STEM TO STREAMS

Expanding upon ideas gained from biomimicry, this next section delves into traditional transdisciplinary learning in educational settings and the possibilities of STREAMS (e.g., Akerson et al., 2018; Mejias et al., 2021). The goals of traditional transdisciplinary learning, particularly STEM education in the United States, focus on integrating disciplines (e.g., science, technology,

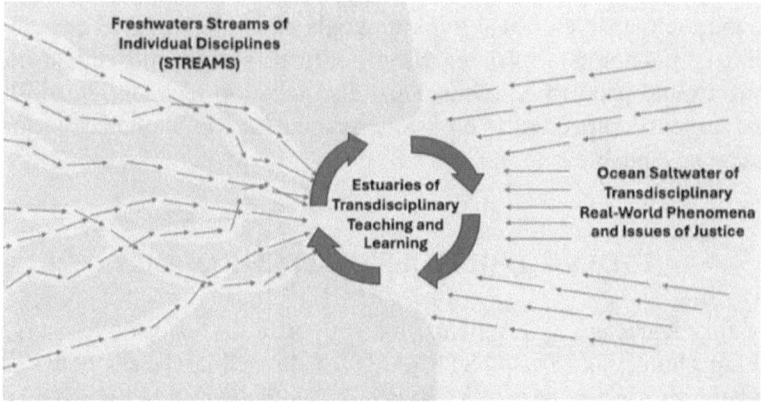

Figure 1.1 Estuaries as a Site for Envisioning STREAMS Teaching and Learning. *Source*: I. Herrick and M. Lawson (2024).

engineering, and mathematics) for students to gain the necessary technical skills to join the STEM workforce (National Research Council, 2011). However, decades of research have shown that a sole focus on technocratic goals limit most students from connecting with or seeing themselves as capable contributors to STEM learning (Aikenhead, 2018). These barriers demand a reimagining of goals through an understanding of why students participate in learning and the inclusion of their voices in how they want to learn and what they want to learn about.

Assisting in this reimagination, STREAMS education calls for the integration of humanistic disciplines, reading/ELA (*R*), and social sciences (*S*), to reorient the purpose and potential outcomes for transdisciplinary learning. Such a reorientation involves moving from problem-solving for technocratic skill-building to problem-solving for humanistic world-building. Through this expansion of what transdisciplinary includes, STREAMS education integrates and values storytelling (reading/ELA) and encourages the use of historical, political, and societal lenses (social sciences) in ways that provide teachers and students opportunities to situate learning and engage in authentic, relevant, and local investigations of phenomena. Through a situated and humanistic approach, STREAMS creates diverse opportunities for students to see themselves as capable problem solvers and to discover active roles they can take in shaping and building their communities' futures. As Morales-Doyle (2024) describes:

> Yes, we need students to apply principles of engineering, when appropriate. But we also need them to recognize when a problem may require a political solution and even more importantly to dream about ways of interacting with each other and the Earth that are not extractive and exploitative. (p. 10)

These more expansive critical thinking goals allow STREAMS education to complexify phenomena by integrating political, social, and ethical considerations around problem-solving. They also develop technocratic skills and nurture students' capacities to envision, imagine, and implement solutions for a just and sustainable future.

RIDING THE TIDES OF TECHNOLOGY

Cultivating learning environments that support a diversity of learners' sensemaking about this *idea* of STREAMS in disciplinary classrooms can be overwhelming for teachers (e.g., Bybee, 2013; Sutherland et al., 2011), particularly when this sensemaking embraces real-world phenomena and issues of justice (e.g., Bartell, 2013; Philip et al., 2016; Redman, 2014). This is due, in part, to the complexity around content knowledge and pedagogical know-how necessary for teachers to cultivate such learning opportunities (Neri et al., 2019). However, for teachers looking to create STREAMS learning, *technology can serve as a practical, flexible, and complementary set of tools and practices to begin mixing disciplines* within classrooms. Drawing upon biomimicry and the phenomenon of an estuary, we see how ocean tides mix freshwater from streams with saltwater. These tidal currents regularly blend fresh and saltwater, creating thriving ecosystems that are rich in biodiversity. In STREAMS education, *riding the tides of technology* can provide classrooms with opportunities to regularly leverage technology to support the blending of ideas to engage in imaginative and critical thinking around a local phenomenon.

Building on the potential of technology in STREAMS education, tools such as images (e.g., pictures and data visualizations) and analytical software (e.g., data dashboards) can help capture and interpret real-world phenomena and justice-related issues. By leveraging these technological assets, classrooms can foster a more integrated understanding and collaborative approach across disciplines, enhancing their ability to address complex societal challenges in a transdisciplinary manner. Furthermore, technology can adapt to the diverse needs and developmental levels of learners. When used effectively, technology can *afford* adaptability and *allow* for a depth of exploration that can be adjusted—much like levels of salinity in the waters of an estuary—providing less salinated shallow areas for beginners to wade through comfortably and deeper, more blended zones for advanced learners to dive into. From this perspective, technology is a tide, dynamically mixing the disciplinary streams of science, reading/ELA, engineering, art, mathematics, and social sciences, enabling students and teachers to engage in transdisciplinary learning that is as dynamic and vibrant as the ecosystems supported by the water in an estuary.

Ellis & colleagues (2020) describe technology as "the tools and practices used by practitioners of science, mathematics, and engineering" (p. 484). We extend this definition to capture the full spectrum of STREAMS disciplines. This broader perspective encompasses the technological tools and practices used in the arts, social sciences, and reading/ELA, highlighting their critical role in enabling a cohesive and inclusive transdisciplinary educational approach. Through this lens, technology extends beyond material devices or digital applications and describes the interplay between material technologies and how these technologies are leveraged, facilitated, and/or implemented. Such a robust definition of technology is warranted because transdisciplinary investigations require concept and skill development with a variety of technologies to fully appreciate and ideate about the complexity of real-world phenomena.

A SHARED TRANSDISCIPLINARY MOMENT: RIDING THE TIDES OF TECHNOLOGY THROUGH COVID-19

The COVID-19 pandemic provided a recent and concrete example of the complexity and issues of justice engaged during transdisciplinary problem-solving and the foundational role of technology across the solutions. In response to the pandemic, scientists across disciplines combined their knowledge and technologies to develop vaccines to combat the virus. However, the reach of the pandemic extended beyond physical well-being and impacted all parts of society, requiring a broader response. Mathematical modeling helped explain the exponential spread of the virus, and data and computer science helped unravel the potential impacts of preventative measures on societal interactions and the economy (e.g., Lee & Campbell, 2020). The arts and reading/ELA supported creative and emotional dimensions of the pandemic vital for individuals and communities to cope with lockdowns and loss (e.g., Chimel et al., 2022; Wan Mak et al., 2021). Historical records were drawn upon to look to the past for how societies previously managed pandemics and contextualize and inspire future possibilities and innovations (e.g., Patterson et al., 2021). Additionally, computers afforded classrooms and other communities the ability to move into online environments for learning and collaboration to continue as best they could. Across all these efforts, the tools and practices of technology contributed to humans making a viable and transdisciplinary response to the pandemic by enabling collaboration, communication, and connection.

During the COVID-19 pandemic, educators encountered an urgent opportunity to foster students' problem-solving skills, engage students in disciplinary practices, and address critical issues of justice (e.g., disproportionate access to medical care) using a transdisciplinary approach. Lee and Campbell (2020) observed these learning opportunities in classrooms and developed an instructional framework drawing on the practical responses of STEM professionals to the pandemic. In this framework, they advocate for the inclusion of data science and computer science in K–12 curricula to help students tackle complex societal problems like COVID-19. Similarly, Lim and colleagues (2022) suggest that using novel forms of data visualization can create new avenues for storytelling with data that bridge and elicit disciplines within a visualization. For example, these authors describe innovative data visualizations, such as an interactive graph of the sound level of New York City streets before and after the pandemic, to describe the pedagogical affordances of these tools to evoke an emotional response and help students think critically about phenomena from multiple perspectives. These educational strategies highlight the shift toward a more integrative teaching approach, where technology facilitates the blending of disciplines, empowering students to make informed, responsible decisions in addressing real-world challenges.

THE CHALLENGES OF NAVIGATING THE TIDES OF TECHNOLOGY

Technologies provide educators and learners with unparalleled access to a myriad of resources, immersive experiences, and the capacity for innovation. However, effectively harnessing technology's full potential demands significant investments of time, effort, and a deep understanding of the ethical, practical, and pedagogical aspects of technology (e.g., Ertmer & Ottenbreit-Leftwich, 2010; Niess et al., 2009). Educators are thus challenged to not only build proficiency with digital tools but also proficiency with integrating these tools into the learning process in ways that expand students' horizons toward broader, more holistic understandings. This means the mere presence of technologies in a classroom does not indicate that teachers or students meaningfully engage with these technologies in ways that support effective STREAMS teaching and learning. For example, a comprehensive study across 47 eighth, ninth, and tenth-grade classrooms in Norway found that despite having national goals for enhancing students' digital skills along with access and infrastructure to support a variety of technologies, teachers' use of technology was narrow and mirrored non-digitized tools (i.e., using digital devices like their non-digitized counterparts of textbooks, worksheets, post-it notes, etc.) (Blikstad-Balas & Klette, 2020). This occurs more broadly as well, as a challenge to successful technology integration is the misalignment between guidelines and teachers' pedagogical and disciplinary beliefs, attitudes, knowledge, and practices (e.g., Ertmer & Ottenbreit-Leftwich, 2010; Ertmer et al., 2012; Koehler & Mishra, 2009; Niess et al., 2009)

We highlight this broad-scale misalignment and shallow integration of technology because the momentum to use technologies in the classroom is increasing (Dubé & Wen, 2022). To better understand and support teachers effectively integrating technology into their practice, Koehler and Mishra (2009) conceptualized teaching with technology as requiring teachers to have substantive technological knowledge that integrates with a teacher's pedagogical content knowledge, forming a technological, pedagogical, and content knowledge for teaching (TPACK). Building on this work, Niess and colleagues (2009) introduced the TPACK Developmental Model as they recognized how teachers' technological and pedagogical content knowledge bases integrated over time as teachers became familiar with and used technologies for teaching. This model highlights how teachers move from *recognizing* the alignment of technologies with learning about their disciplinary field to becoming *accepting* of teaching their disciplinary field through technologies and finally toward *adapting*, *exploring*, and *advancing* the ways in which technologies are utilized in day-to-day classroom teaching and learning (Niess et al., 2009). It is important to note that due to the variety of technologies available, teachers often experience these developmental stages in cycles, as they continuously encounter and discern the validity of newly available technologies. This ongoing and cyclical process further explicates the challenge of aligning guidelines for technology use and teachers' pedagogical and disciplinary visions for instruction.

RIDING THE TIDES OF TECHNOLOGY IN PERMANENT WHITE WATER

Recognizing the promise and perils of technology for providing opportunities to invite transdisciplinary learning, more attention should be paid to understanding and supporting teachers to navigate these challenges. To begin, many of the complexities teachers face stem from pressures related to high-stakes testing, a lack of understanding of flexible classroom practices, gaps in content knowledge, prescriptive routines, and teachers' sociocultural agency (e.g., Freire & Valdez, 2017; Goldenberg, 2014; Moll et al., 1992; Warren, 2014). Collectively, these issues create barriers for teachers to develop justice-centered learning opportunities (e.g., Morales-Doyle, 2017) and incorporate practical ways to engage transdisciplinary learning into their pedagogical visions for instruction (Borrero & Sanchez, 2017; Vaughn, 2021). The tensions around technology integration, pedagogical moves that center students, inviting conversations about justice, and the standards that guide curriculum embody the conditions of *permanent white water* teachers persistently must navigate. Vaill (1996) used the term "permanent white water" to describe how organizational leaders navigate the turbulence and uncertainty that stems from "surprising, novel, messy, costly, and unpreventable" events (p. 14). He describes that navigating these waters requires leaders who can learn and adapt to their present conditions while exercising their will and judgment in macro- and micro-systems. These conditions of permanent white water extend to classrooms, where teachers also face continuous and unpredictable challenges (Wergin, 2019).

As we consider how transdisciplinary classrooms mimic an estuary, we must recognize the increased white water generated from the mixing of fresh and saltwater that persists when exploring phenomena in transdisciplinary ways. These conditions require that teachers adapt and learn while responding to novel and diverse situations that arise. In the context of STEM education, teachers who are skilled at supporting transdisciplinary explorations recognize the need to incorporate conversations about justice and integrate themes of privilege, diversity, and sociohistorical inequities into these explorations (Crowley, 2016; Emdin, 2016; Ladson-Billings, 2024). These educators utilize a healing-informed approach to teaching and learning that holds space for students' emotions to be shared and processed by engaging in disciplinary practices while examining systemic injustices contributing to the creation of disproportionate access to opportunity (e.g., Kokka, 2019). Through this approach, they view students as knowledgeable with valuable out-of-school experiences that contribute significantly to their understanding of STEM content and inform their desire to engage in disciplinary practices (e.g., Calabrese Barton et al., 2020; Kokka, 2019; Herrick, 2023). In the context of STREAMS education, teachers need to expand how they view students' out-of-school knowledge and experiences to include their emotions and ideas connected to broader parts of their worlds (e.g., social science lenses). Teachers beginning to work in these ways must develop reflexes, like white water river guides, that support them in responding quickly and appropriately to each incoming obstacle and wave (Herrick, 2023).

TWO JUSTICE-CENTERED FLEXIBLE ROUTINES FOR STREAMS EDUCATION

Becoming a successful navigator of transdisciplinary permanent white water requires teachers to stretch and grow in developmentally appropriate ways (Drago-Severson & Blum-DeStefano, 2017; Herrick, 2023). In the following sections, we highlight two *justice-centered flexible routines* that leverage effective use of technology in practical and powerful ways and provide an entry point for teachers to begin navigating and developing expertise in these waters: (1) Photovoice and (2) Community Science Data Talks. In *photovoice*, the technology of a camera provides a means for students to bring their out-of-school understandings of phenomena as part of classroom learning. In *Community Science Data Talks*, the technology of data dashboards and data visualizations provide a platform for students to interrogate data about local transdisciplinary phenomena. These routines can be taken up by teachers and students in any grade-level or disciplinary classroom and move their collective work toward transdisciplinary investigations. Additionally, as classrooms continue to engage with these routines, we have found that students take more control and autonomy for their learning over time. As technology assists students in making sense of phenomena and connecting disciplines in novel ways, students then take larger roles by configuring and reconfiguring classroom discourse and future investigations. We unpack what these routines are and what we have learned from researching these routines alongside the teachers and students below.

PHOTOVOICE

What Is Photovoice?

Photovoice is a justice-centered research methodology typically used in community-based participatory action research projects (e.g., Comas-Díaz & Rivera, 2020; Rhodes et al., 2015; Wang & Burris, 1997). This method distinguishes itself from traditional research because it intentionally shifts the power dynamic from a researcher studying participants *to* community members doing research by documenting and sharing their own experiences (Comas-Díaz & Rivera, 2020; Rhodes et al., 2015; Wang & Burris, 1997). Photovoice adopts a tiered approach that begins with individuals taking photos to document some event/phenomena and then progressively reflecting on and sharing these photos with groups and communities to influence policy and drive social change (Comas-Díaz & Torres Rivera, 2020). At its core, photovoice seeks to highlight the significance of individual and shared knowledge and experiences within communities and to emphasize the importance of empowering community voices in research.

The justice-oriented goals of empowerment and positioning of participants as experts extend to bringing photovoice into the classroom as a pedagogical routine. In the classroom, photovoice encourages a similar power dynamic shift as described above, but between teachers and students. Here, students use the technology of a camera (e.g., disposable or digital camera, smartphone, tablet) to drive inquiry into how *they* see and experience transdisciplinary issues in their everyday lives. For example, we asked fifth graders to take pictures of how they see and/or experience the impacts of climate change in their everyday lives (Herrick et al., 2022). In this research, students utilized the technology of a camera in ways that positioned them as experts on local climate issues within their community and as capable documentarians with valuable stories to share. Additionally, teachers provided students with guided reflection prompts (e.g., What do you see here? How does it impact our lives? Why does this asset or problem exist?) to support students in reseeing their images and connecting these ideas to broader content knowledge and engaging disciplinary practices. Towards these efforts, incorporating technology and the arts in science learning provides students with different ways to communicate their thinking, emotions, and personal genius about concepts and phenomena (Chandler et al., 2020; Emdin, 2021).

What We Found

Photovoice positioned fifth-grade students as qualitative researchers who collected valuable data through their photographs and engaged in careful reflection on these images to reveal deeper and more nuanced understandings of their worlds. Through this process, students began to authentically share their *funds of* knowledge—the cultural and familial knowledge and practices embedded in students' out-of-school lives (Moll et al., 1992)—and *funds of* feeling—the more complex emotions, where cognition and emotion are tied to feelings about the places they inhabit and shape how new information is processed (Levine, 2021). In small groups, students began to organize these emerging ideas and emotions by categorizing their photographs to document how *they collectively* see climate change in their everyday lives. At the end of the project, students shared their findings with the whole class and collaborated on identifying patterns across the data corpus.

The photographs showcased various perspectives, knowledge, and emotions each student had about climate change's impacts on their community. Through reflections and interviews, we found students progressively incorporated scientific language in their descriptions and explanations of their photos (Herrick et al., 2022). This progression underscored students' growing confidence in their content knowledge, ability to use scientific language, and ability to apply their knowledge to make sense of the impacts of climate change and support their communities. For example, students recognized the collective responsibility of humans in exacerbating climate change and expressed ideas on collective actions to mitigate its effects (e.g., selecting a month every year to be a vegetarian to support sustainability causes). Students also demonstrated an awareness of the role of time in climate change, comparing past experiences with current situations. Additionally, in a whole-class conversation after their presentations, we saw students' emotions shift from feelings of hopelessness about climate change to a sense of protectiveness and urgency for their communities. Through this urgency, students engaged in self-driven investigations into constructing ideas around what they could do to support their local community in mitigating the impacts of climate change (e.g., educating community members to raise awareness of the impacts). Overall, we found photovoice to be an effective justice-centered flexible routine that can support students and teachers with engaging in locally relevant transdisciplinary topics while supporting students in constructing hope (Ojala, 2012a) for their futures through finding desired actions they can take in their local community (Herrick et al., 2022).

COMMUNITY SCIENCE DATA TALKS

What Are Community Science Data Talks?

Recent developments in data science, particularly increased access for students to interact with and create data, present unique opportunities for all classrooms. These advancements enable teachers to employ data as a transdisciplinary tool, aiding students in developing data literacy skills (e.g., analyzing and interpreting, modeling with, or asking critical questions of data) that enhance their depth of learning and ability to participate in civic activities (Herrick, 2023). Further, using local data allows students to gain a deeper understanding of their communities and empowers them to tell stories or counter-narratives that can develop everyone's understanding of the complex, multi-layered nature of phenomena and how communities come to be (e.g., Kokka, 2020; Louie et al., 2021; Rubel et al., 2016; Taylor et al., 2020; Van Wart et al., 2020). These experiences with local data in the classroom can also foster strong emotional responses, such as negative emotions like anxiety and worry or more positive emotions like pride, hope, and action. This requires a healing-informed approach where students identify and share the systemic injustices that create the complex issues facing communities in order to support students in productively processing these emotions toward feelings of empowerment and hope (Kokka, 2019; Ojala, 2023).

Engaging issues of justice in the classroom can be intimidating for teachers and students, as much of this work occurs through time and resource-intensive large-scale projects that engage multiple partnerships (e.g., Bang et al., 2016; Morales-Doyle & Frausto, 2021). Thus, we developed Community Science Data Talks (CSDTs), a small-scale version of this work that could serve as an entry point for classrooms to begin engaging in justice-centered STREAMS. Theoretically and pedagogically, CSDTs are informed by critical place-based education (e.g., Gruenwald, 2003), transformative experience (e.g., Pugh, 2011), and justice-centered STEM (e.g., Gutstein, 2006, 2012; Morales-Doyle, 2017) (see figure 1.2). This grounding for CSDTs also necessitates what Kokka (2022) describes as teachers planning for and attending to dominant, critical, and affective pedagogical goals when discussing issues of justice in the classroom. Here, *dominant pedagogical goals* represent the disciplinary standards associated with the phenomena students are learning about, *critical pedagogical goals* represent the development of students' sociopolitical consciousness in relation to the phenomena and impacts under investigation, and *affective pedagogical goals* recognize the inherent emotionality of social justice science issues and intend to support students in productively processing their emotions. At the intersection of this framework and these pedagogical goals, CSDTs intend for teachers to move student

Figure 1.2 Theoretical and Pedagogical Framework of Community Science Data Talks.
Source: I. Herrick and M. Lawson (2024).

discourse from reaction to action, where the culmination of a series of CSDTs pushes student conversations beyond the implications of data and encourages them to reimagine what the data story *could* or *should* tell.

In practice, CSDTs are a 10–15 minute habitual discourse routine prompted by a local data visualization that brings conversations about local issues of justice into classrooms. Over time, students gain a more holistic understanding of relationships between phenomena, justice, and place as data visualizations highlight the necessity for a variety of lenses during investigations (e.g., applying historical, environmental, and socioeconomic lenses). The aim of CSDTs is to create a learning environment that is not only justice-centered but also emotionally aware, equipping students with the cognitive and emotional tools to navigate and impact their world positively. Students are also encouraged to share their out-of-school funds of knowledge (i.e., family, community, and cultural knowledge; Moll et al., 1992) and funds of feeling (i.e., complex emotions tied to feelings about place; Levine, 2021) just as much as their disciplinary knowledge during a CSDT. This holistic approach underscores the importance of nurturing intellectual and emotional competencies in students, preparing them for the complexities of navigating the world.

CSDTs are not intended to be a one-and-done routine but rather a habitual routine (e.g., every Monday) in which classrooms engage. This is because complex phenomena and issues of justice, such as the disproportionate impacts of climate change on communities, cannot be unpacked in a single conversation. Instead, a single CSDT provides a foundation for future investigations where phenomena and issues of justice are unpacked in meaningful ways by layering different lenses on the same phenomena and place (e.g., historical lens, socioeconomic lens, environmental lens). By habitually engaging with CSDTs, students can make sense of and see their place through multiple lenses and build a nuanced comprehension of phenomena and their impact on their community. Finally, as teachers plan for the dominant, critical, and affective pedagogical goals, they must also plan questions that support students in making progress toward these goals. These planned questions could include "What patterns do you see in the data?" (dominant), "Who is impacted by this data?" (critical), and "How do these data make you feel?" (affective).

What We Found

During three iterations of design-based research (Cobb et al., 2003; Sandoval & Bell, 2010) working with 15 K–12 classrooms in North and South America, we found that CSDTs activated transdisciplinary learning using local data visualizations that depict real-world phenomena and issues of climate justice (Herrick, 2023; Herrick et al., in press). During these conversations, students shared their individual and collective knowledge—both experiential and emotional—which opened new dimensions in the classroom to think with and discuss. These discussions often crossed disciplines and engaged multiple perspectives (e.g., scientific, mathematical, historical, environmental, sociocultural, and socioeconomic) as students interrogated the data to make sense of climate justice issues affecting their communities (Herrick, 2023). Drawing on the *idea* of STREAMS in CSDTs encouraged students to think critically and discuss how data related to their lives and communities, thus embodying the essence of STREAMS through the movements and mergers of students' emotions with transdisciplinary inquiries. At the same time, teachers were able to stretch their thinking about teaching and learning in justice-oriented ways (Herrick, 2023).

We also found that emotions acted as a catalyst for deeper engagement in various disciplinary practices (Herrick, 2023). As students shared their funds of knowledge (Moll et al., 1992) and funds of feeling (Levine, 2021), these contributions entangled into a contagious process where students wanted to find solutions for local issues. Here, their dialogue configured and reconfigured discourse in ways that supported students to construct hope for their futures (Ojala, 2012a). For an example of this configuring and reconfiguring, we observed how two secondary classrooms dynamically arrived at the same emotional configuration (i.e., Vea, 2020) of constructive hope after a series of CSDTs. In this example, both classrooms were located in a large urban city in the southwestern region of the United States; however, these classrooms were at different schools located in different areas of the city. One classroom population had a majority of white students from families with high socioeconomic status (i.e., experiencing privilege), while the other classroom had a majority population of students from families with low socioeconomic status with backgrounds that have been historically minoritized (Herrick, 2023). The students experiencing privilege constructed hope through the emotional pathway of critical civic empathy (Mirra, 2018), while the students from historically minoritized backgrounds constructed hope through the emotional pathway of communal value (Gray et al., 2020). These different pathways toward hope emerged from the intersectional and contextual aspects of privilege and oppression students experienced as a result of their identities and how these aspects configured and reconfigured each other over the discussion.

Overall, we found CSDTs to be an effective justice-centered flexible routine that can support students and teachers in making sense of complex phenomena in transdisciplinary ways over time. The habitual nature of CSDTs supported students' productive processing of emotion as the teachers consistently planned for cognitive and emotional dimensions of learning through dominant, critical, and affective pedagogical goals. Additionally, as teachers continued engaging in this routine, they began to better understand how to navigate and maneuver in the transdisciplinary permanent white water.

CONCLUSIONS

Through Photovoice and Community Science Data Talks, we share how the implementation of *justice-centered flexible routines* supported the movement and merger of students' agency and autonomy to engage in self-determined actions toward constructing more just and sustainable futures (Herrick et al., 2022; Herrick, 2023). These routines also supported the movement and merger of teachers developing more complex orientations around their practices and purposes for teaching and learning. It is through these studies we demonstrate how exploratory technologies and small-scale *justice-centered flexible routines* can create opportunities for STREAMS to emerge and propagate.

As an *idea*, STREAMS education can symbolize what is possible in transdisciplinary teaching and learning. Through the lens of an estuary, we can imagine how engaging students with the *idea* of STREAMS can foster adaptable learning environments that support situated and humanistic transdisciplinary investigations (e.g., Herrick, 2023; Kokka, 2022; Morales-Doyle, 2017). We are also able to imagine how blending disciplines with locally relevant phenomena breeds vibrant and diverse ways of thinking, akin to fresh and saltwater blending together in an estuary to support rich, biodiverse ecosystems. However, as with any *idea* there are endless possibilities and situational constraints.

Situational climates threaten and constrain both the *idea* of STREAMS education and the vibrancy of an estuary. Climate change, for example, threatens and constrains the biodiversity of estuaries in the Mississippi Delta and the Chesapeake Bay, and the climate of permanent white water in U.S. schools threatens the idea of STREAMS coming to life. Thus, this chapter explores how, within the climate of permanent white water, teachers can ride the tides of technology to ignite STREAMS opportunities in practical and small-scale ways. Doing so allows teachers to develop navigation skills for transdisciplinary permanent white water and supports students in problem-solving, knowledge-building, and discovering opportunities to act toward building more just and sustainable futures.

REFERENCES

Aikenhead, G. S. (2006). *Science education for everyday life: Evidence-based practice*. Teachers College Press.

Akerson, V. L., Burgess, A., Gerber, A., Guo, M., Khan, T. A., & Newman, S. (2018). Disentangling the meaning of STEM: Implications for science Education and science teacher education. *Journal of Science Teacher Education, 29*(1), 1–8.

Bang, M., Faber, L., Gurneau, J., Marin, A., & Soto, C. (2016). Community-based design research: Learning across generations and strategic transformations of institutional relations toward axiological innovations. *Mind, Culture, and Activity, 23*(1), 28–41.

Barraza, L. (1999). Children's drawings about the environment. *Environmental Education Research, 5*(1), 49–66.

Bartell, T. G. (2013). Learning to teach mathematics for social justice: Negotiating social justice and mathematical goals. *Journal for Research in Mathematics Education, 44*(1), 129–163.

Blikstad-Balas, M., & Klette, K. (2020). Still a long way to go: Narrow and transmissive use of technology in the classroom. *Nordic Journal of Digital Literacy, 15*(1), 55–68.

Borrero, N., & Sanchez, G. (2017). Enacting culturally relevant pedagogy: Asset mapping in urban classrooms. *Teaching Education, 28*(3), 279–295.

Bybee, R. W. (2013). *The case for STEM education: Challenges and opportunities.* National Science Teachers Association Press.

Calabrese Barton, A., Tan, E., & Birmingham, D. J. (2020). Rethinking high-leverage practices in justice-oriented ways. *Journal of Teacher Education, 7*(14), 477–494.

Chandler, P., Osnes, B., & Boykoff, M. (2020). Creative climate communications: Teaching from the heart through the arts. In J. Henderson & A. Drewes (Eds.), *Teaching Climate Change in the United States* (pp. 172–185). Routledge.

Chmiel, A., Kiernan, F., Garrido, S., Lensen, S., Hickey, M., & Davidson, J. W. (2022). Creativity in lockdown: Understanding how music and the arts supported mental health during the COVID-19 pandemic by age group. *Frontiers in Psychology, 13*:993259, 1–18.

Cobb, P., Confrey, J., diSessa, A., Lehrer, R., & Schauble, L. (2003). Design experiments in educational research. *Educational Researcher, 32*(1), 9–13.

Comas-Díaz, L., & Torres Rivera, E. (Eds.). (2020). *Liberation psychology: Theory, method, practice, and social justice.* American Psychological Association.

Crowley, R. M. (2016). Transgressive and negotiated White racial knowledge. *International Journal of Qualitative Studies in Education, 29*(8), 1016–1029.

Drago-Severson, E., & Blum-DeStefano, J. (2017). The self in social justice: A developmental lens on race, identity, and transformation. *Harvard Educational Review, 87*(4), 457–481.

Dubé, A. K., & Wen, R. (2022). Identification and evaluation of technology trends in K12 education from 2011 to 2021. *Education and information technologies, 27*(2), 1929–1958.

Ellis, J., Wieselmann, J., Sivaraj, R., Roehrig, G., Dare, E., & Ring-Whalen, E. (2020). Toward a productive definition of technology in science and STEM education. *Contemporary issues in technology and teacher education, 20*(3), 472–496.

Emdin, C. (2016). *For White folks who teach in the hood . . . and the rest of y'all too: Reality pedagogy and urban education.* Beacon Press.

Emdin, C. (2021). *Reimagining the Culture of Science, Technology, Engineering, and Mathematics Stem, Steam, Make, Dream.* International Center for Leadership in Education, Inc.

English, L. D. (2016). STEM education K-12: perspectives on integration. *International Journal of STEM Education, 3*(1), 3.

Ertmer, P. A., & Ottenbreit-Leftwich, A. T. (2010). Teacher technology change: How knowledge, confidence, beliefs, and culture intersect. *Journal of research on Technology in Education, 42*(3), 255–284.

Ertmer, P. A., Ottenbreit-Leftwich, A. T., Sadik, O., Sendurur, E., & Sendurur, P. (2012). Teacher beliefs and technology integration practices: A critical relationship. *Computers & education, 59*(2), 423–435.

Freire, J.A., & Valdez, V.E. (2017). Dual language teachers' stated barriers to implementation of culturally relevant pedagogy. *Bilingual Research Journal, 40*(1), 55–69.

Goldenberg, B.M. (2014). White teachers in urban classrooms: Embracing non-white students' cultural capital for better teaching and learning. *Urban Education, 49*(1), 111–144.

Gray, D.L., McElveen, T.L., Green, B.P., & Bryant, L.H. (2020). Engaging black and latinx students through communal learning opportunities: A relevance intervention for middle schoolers in STEM elective classrooms. *Contemporary Educational Psychology, 60*, 101833.

Gruenewald, D. A. (2003). Foundations of place: A multidisciplinary framework for place-conscious education. *American Educational Research Journal, 40*(3), 619–654.

Gutiérrez, R. (2008). A "gap-gazing" fetish in mathematics education? Problematizing research on the achievement gap. *Journal for Research in Mathematics Education, 29*(4), 357–364.

Gutstein, E. (2006). *Reading and writing the world with mathematics: Toward a pedagogy for social justice.* Taylor & Francis.

Gutstein, E. (2012). Connecting community, critical, and classical knowledge in teaching mathematics for social justice. In S. Mukhopadhyay & W.-M. Roth (Eds.), *Alternative Forms of Knowing (in) Mathematics* (pp. 299–311). Sense Publishers.

Herrick, I. R. (2023). What do you notice? What do you wonder? A mixed-methods investigation into community science data talks [Doctoral Dissertation, University of Southern California]. ProQuest One Academic.

Herrick, I. R., Lawson, M., & Matewos, A. M. (2022). Through the eyes of a child: Exploring and engaging elementary students' climate conceptions through photovoice. *Educational and Developmental Psychologist, 39*(1), 100–115.

Herrick, I. R., Lawson, M., & Matewos, A. M. (in press). *"How do these data make you feel?"*: The emergence of emotional pathways in community science data talks about climate justice issues. *Science Education.*

Hickman, C., Marks, E., Pihkala, P., Clayton, S., Lewandowski, E., Mayall, E. E., Wray, B., Mellor, C., & van Susteren, L. (2021). Climate anxiety in children and young people and their beliefs about government responses to climate change: A global survey. *The Lancet Planetary Health, 5*(12), 863–873.

Immordino-Yang, M. H., Darling-Hammond, L., & Krone, C. R. (2019). Nurturing Nature: How Brain Development Is Inherently Social and Emotional, and What This Means for Education. Educational Psychologist, 54(3), 185–204.

Kelley, T. R., & Knowles, J. G. (2016). A Conceptual framework for integrated STEM education. *International Journal of STEM Education, 3*(11), 1–11.

Koehler, M., & Mishra, P. (2009). What is technological pedagogical content knowledge (TPACK)? *Contemporary Issues in Technology and Teacher Education, 9*(1), 60–70.

Kokka, K. (2019). Healing-informed social justice mathematics: Promoting students' sociopolitical consciousness and well-being in mathematics class. *Urban Education, 54*(9), 1179–1209.

Kokka, K. (2020). Social Justice Pedagogy for Whom? Developing Privileged Students' Critical Mathematics Consciousness. The Urban Review, 52(4), 778–803.

Kokka, K. (2022). Toward a theory of affective social pedagogical goals for social justice mathematics. *Journal for Research in Mathematics Education, 53*(2), 133–153.

Ladson-Billings, G. (2024). *Justice matters.* Bloomsbury Academic.

Lee, O., & Campbell, T. (2020). What science and STEM teachers can learn from COVID-19: Harnessing data science and computer science through the convergence of multiple STEM subjects. *Journal of Science Teacher Education, 31*(8), 932–944.

Levin, S. (2021). *Up-down-both-why: A funds of feeling approach to literature. Cult of Pedagogy.*

Lim, V. Y., Peralta, L. M. M., Rubel, L. H., Jiang, S., Kahn, J. B., & Herbel-Eisenmann, B. (2022). Keeping pace with innovations in data visualizations: A commentary for mathematics education in times of crisis. *ZDM Mathematics Education, 55*, 109–118.

Louie, J., Stiles, J., Fagan, E., Roy, S., & Chance, B. (2021). Data investigations to further social justice inside and outside of STEM. *Connected Science Learning, 3*(1).

Mak, H. W., Fluharty, M., & Fancourt, D. (2021). Predictors and impact of arts engagement during the COVID-19 pandemic: Analyses of data from 19,384 adults in the COVID-19 social study. *Frontiers in Psychology, 12*:626263, 1–17.

Mejias, S., Thompson, N., Sedas, R. M., Rosin, M., Soep, E., Peppler, K., Roche, J., Wong, J., Hurley, M., Bell, P., & Bevan, B. (2021). The trouble with STEAM and why we use it anyway. *Science Education, 105*(2), 209–231.

Mirra, N. (2018). *Educating for Empathy: Literacy Learning and Civic Engagement.* Teachers College Press.

Moll, L., Amanti, C., Neff, D., & Gonzalez, N. (1992). Funds of knowledge for teaching: Using a qualitative approach to connect homes and classrooms. *Theory into Practice, 31*(2), 132–141.

Morales-Doyle, D. (2017). Students as curriculum critics: Standpoints with respect to relevance, goals, and science. *Journal of Research in Science Teaching, 55*(5), 749–773.

Morales-Doyle, D. (2024). *Transformative Science Teaching: A Catalyst for Justice and Sustainability. United States*: Harvard Education Press.

Morales-Doyle, D., & Frausto, A. (2021). Youth participatory science: a grassroots science curriculum framework. *Educational Action Research, 29*(1), 60–78.

National Research Council (2011). *Successful K-12 STEM Education: Identifying Effective Approaches in Science, Technology, Engineering, and Mathematics.* The National Academies Press.

Neri, R. C., Lozano, M., & Gomez, L. M. (2019). (Re)framing resistance to culturally relevant education as a multilevel learning problem. *Review of Research in Education, 43*(1), 197–226.

Niess, M. L., Ronau, R. N., Shafer, K. G., Driskell, S. O., Harper, S. R., Johnston, C., Browning, C., Özgün-Koca, S. A. & Kersaint, G. (2009). Mathematics teacher TPACK standards and development model. *Contemporary Issues in Technology and Teacher Education, 9*(1), 4–24.

Noddings, N. (2002). *Starting at Home: Caring and Social Policy.* Univ of California Press.

Ojala, M. (2012a). Hope and climate change: The importance of hope for environmental engagement among young people. *Environmental Education Research, 18*(5), 625–642.

Ojala, M. (2012b). Regulating worry, promoting hope: How do children, adolescents, and young adults cope with climate change? *International Journal of Environmental and Science Education, 7*(4), 537–561.

Ojala, M. (2015). Hope in the face of climate change: Associations with environmental engagement and student perceptions of teachers' emotion communication style and future orientation. *Journal of Environmental Education, 46*(3), 1–16. https://doi.org/10.1080/00958 964.2015.1021662

Ojala, M. (2016). Young people and global climate change: Emotions, coping, and engagement in everyday life. In N. Ansell, N. Klocker, & T. Skelton (Eds.), Geographies of global issues: Change and threat: Geographies of children and young people (Vol. 8, pp. 1–19). Singapore: Springer. https://doi.org/10.1080/00958964.2015.1021662.

Ojala, M. (2023). Climate change education and critical emotional awareness (CEA): Implications for teacher education. *Educational Philosophy and Theory, 55*(10), 1109–1120.

Patterson, G. E., McIntyre, K. M., Clough, H. E., & Rushton, J. (2021). Societal impacts of pandemics: Comparing COVID-19 with history to focus our response. *Frontiers in Public Health, 12*:630499, 1–6.

Philip, T. M., Olivares-Pasillas, M. C., & Rocha, J. (2016). Becoming Racially Literate About Data and Data-Literate About Race: Data Visualizations in the Classroom as a Site of Racial-Ideological Micro-Contestations. *Cognition and Instruction, 34*(4), 361–388.

Pugh, K.J. (2011). Transformative experience: An integrative construct in the spirit of Deweyan pragmatism. *Educational Psychologist, 46*(2), 107–121.

Redman, E.H. (2014). *A study of novice science teachers' conceptualizations of culturally relevant pedagogy.* Doctoral Dissertation, University of California, Los Angeles.

Rhodes, S. D., Alonzo, J., Mann, L., M. Simán, F., Garcia, M., Abraham, C., & Sun, C. J. (2015). Using photovoice, Latina transgender women identify priorities in a new immigrant-destination state. *International Journal of Transgenderism, 16*(2), 80–96.

Rubel, L. H., Lim, V. Y., Hall-Wieckert, M., & Sullivan, M. (2016). Teaching mathematics for spatial justice: An investigation of the lottery. *Cognition and Instruction, 34*(1), 1–26.

Sandoval, W. A., & Bell, P. (2004). Design-based research methods for studying learning in context: Introduction. *Educational psychologist, 39*(4), 199–201.

Sherin, M., & van Es, E. A. (2009). Effects of video club participation on teachers' professional vision. *Journal of teacher education, 60*(1), 20–37. Shudak, N. J., & Avoseh, M. B. (2015). Freirean-based critical pedagogy: The challenges of limit-situations and critical transitivity. *Creative Education, 6*(04), 463.

Southerland S. A., Sowell S., Blanchard M., Granger E. M. (2011). Exploring the construct of pedagogical discontentment: A tool to understand science teachers' openness to reform. *Research in Science Education, 41*, 299–317.

Taylor, S., Landry, C. A., Paluszek, M. M., Fergus, T. A., McKay, D., & Asmundson, G. J. (2020). COVID stress syndrome: Concept, structure, and correlates. *Depression and Anxiety, 37*(8), 706–714.

Vaill, P. B. (1996). *Learning as a way of being: Strategies for survival in a world of permanent white water.* Jossey-Bass.

Van Wart, S., Lanouette K., & Parikh, T. S. (2020). Scripts and counterscripts in community-based data science: Participatory digital mapping and the pursuit of a Third Space. *Journal of the Learning Sciences, 20*(1), 127–153.

Wang, C., & Burris, M. A. (1997). Photovoice: Concept, methodology, and use for participatory needs assessment. *Health Education & Behavior, 24*(3), 369–387.

Warren, C. A. (2014). Towards a pedagogy for the application of empathy in culturally diverse classrooms. *The Urban Review, 46*, 395–419.

Wergin, J. F. (2019). *Deep learning in a disorienting world.* Cambridge University Press.

Chapter 2

Digital Humanities and STREAMS

Pre-service Teachers' Exploration of Innovative Curricular Pathways

Britta Bletscher and Heidi L. Hallman

This chapter considers how the subject of English language arts, particularly with its increased focus on technology, is a part of STREAMS. Examining the way that pre-service English teachers learn to use digital humanities as a technology and pedagogical tool in a technology for teachers course, this chapter highlights specific pedagogies that pre-service English teachers might use in their future classrooms.

Digital humanities (DH) is an area of scholarly activity at the intersection of computing or digital technologies and the disciplines of the humanities. It is particularly well-suited to help pre-service teachers see technology integration into English language arts as producing a new kind of knowledge in the teaching of English–knowledge that could not be produced without the use of information technology. DH started as a digitization and computing practice that created an "environment in which physical and virtual realms merge in fluid and seamless ways" (Jones, 2016, p. 4). These practices used investigation, analysis, and synthesis to present data in electronic form. Educators quickly saw the value in DH, as it allows for increased accessibility, depth, and awareness of how media can directly impact the humanities.

As DH grew in popularity, its uses began to evolve. At its core, the relationship to data, analysis, and media remained the same, but the scale at which these concepts were performed grew. While still relying on the earlier uses of computing, "the new model of digital humanities emphasized . . . the analysis and visualization of *large datasets of humanities materials*" which included "'distant reading,' engaged in coding and building digital tools and websites and archives as well as wearable processors and other devices, and responded to the 'spatial turn' across the disciplines with data-layered 'thick mapping' projects" (Jones, 2016, p. 4, emphasis added). These projects encompass more data, bigger connections, and larger relevance; incorporating them into classrooms helps students to see the interdisciplinarity of different ideas and to think more critically about how various subjects interweave. The current state of DH puts the "digital into reciprocal conversations with an array of cultural artifacts, the objects on which humanistic study has historically been based and new kinds of objects, including born-digital artifacts" (Jones, 2016, p. 5). This exposure makes education more interesting and relevant, and when integrated into ELA courses, invests students more thoroughly in the learning material and opens passageways to understanding English's larger applicability.

Through studying pre-service teachers' digital humanities projects, created as a component of a technology for teachers course, this chapter highlights the discipline of English language arts as one that goes beyond just reading and writing and instead draws upon critical inquiry alongside the use of technology.

THE CONTEXT: TECHNOLOGY FOR TEACHERS COURSE

The teacher education program at Green State University (all names of people and places are pseudonyms) includes a course, EDUC 550: Technology for Teachers. Green State University is the state's flagship institution, a large, research-oriented university in the Midwest United States. All pre-service teachers in Green State's elementary and secondary teacher education programs enroll in the course. The particular course that was the site of inquiry for this chapter is the EDUC 550 course for pre-service teachers in the secondary education disciplines of English Language Arts (ELA), History and Government, and Foreign Language.

EDUC 550 focuses on the ISTE standards for educators, which include (a) facilitating and inspiring student learning and creativity, (b) designing and developing digital-age learning experiences and assessments, and (c) modeling digital-age work and learning. Pre-service teachers in the course are engaged with hands-on experiences with various information and educational technologies, and learn how to use them to develop technology-enriched learning materials. Digital humanities (DH) is one of the critical technology components of the course, especially for teachers in the discipline of ELA.

The remainder of the chapter discusses specific examples of pre-service teachers' exploration of digital humanities (DH). Three sections: (1) Knightlab to introduce digital humanities, (2) FilterBubbles to bridge digital humanities and social media, and (3) Voyant for text analysis, illustrate the range of how DH can be used as part of the English language arts curriculum.

KnightLab as an Introduction to Digital Humanities

As part of the Digital Humanities unit in EDUC 550, students were asked to create an interactive map on KnightLab, a website devoted to innovating digital journalism. KnightLab was created by designers, developers, educators, and students from Northwestern University with the intention of establishing a "collaborative environment for interdisciplinary exploration and conversation" (Northwestern University, 2023). The KnightLab website allows users to stay up to date on current research and news, look at sample projects, and create their own timelines, storylines, juxtaposition of scenes, and storymaps. KnightLab is an excellent way to introduce students to DH projects, as its processes are fairly straightforward and easy to navigate. To create a storymap, students enter the names of the locations in the order they want them to appear, then add titles, descriptions, and photos for each location. Once complete, students should be able to virtually travel through their story. This piece of DH allows for a narrative to be represented digitally.

Because of the interdisciplinarity of a storymap, class conversations are able to expand beyond the use of technology in the classroom. Students learn how to read map coordinates, understand maps more generally, and discover aspects of geography that were previously unknown. In addition, since DH relies heavily on ethics and credibility, conversations can be had about the importance of copyright and how copyright laws tie into DH work more holistically.

After spending some time working with the software in class, students acknowledged its benefits and drawbacks and brainstormed how they might use it in their own future assignments and curricula. The wide breadth of knowledge that goes into creating a storymap makes this assignment be more applicable to students than other assignments in the course. It increases awareness of unique topics such as reading maps and establishing credibility, while also allowing space for students to discover how this piece of DH fits into their own teaching spaces. These takeaways are illustrated through the student storymap assignment submissions and reflections.

Students utilized KnightLab to create maps that depict a wide array of personal, fictional, and historical narratives, showing that DH software is flexible and can be adapted to fit learning goals and unit objectives for many subjects, particularly English Language Arts. One student, Taylor, created a map titled, "All of the Amazing Places Taylor Went Over the Summer," which included all the stops they made throughout their summer road trip. Taylor's storymap takes her audience through the trip chronologically, with a fun title, a short description of their activities, and a personal photo of their time for each location. This personal depiction shows how KnightLab can be used to help students illustrate a story chronologically, visually, and descriptively. It requires recall and is a great way to demonstrate understanding of any type of narrative. Another student, Micah, created a storymap titled, "The Last of Us: Joel and Ellie's Journey" (see figure 2.1). This map takes its audience through the happenings of the first season of HBO's hit series, *The Last of Us*. In the eight-episode debut, the main characters, Joel and Ellie, battle their way across America after an apocalyptic pandemic. Micah depicts this traumatizing expedition through their use of locations they visited, titles and short descriptions, and copyright-approved photos. This storymap shows how KnightLab can be used to keep track of intricate plots, similar settings, and countless secondary characters. Regardless of the type of narrative, creating a storymap has many curricular benefits.

In addition to submitting a storymap, students wrote reflections about their experiences using the DH software and how they envisioned it being used in future curriculum. When considering the advantages, Taylor noted, "I believe a benefit . . . is becoming more familiar with a map. I do not do much work

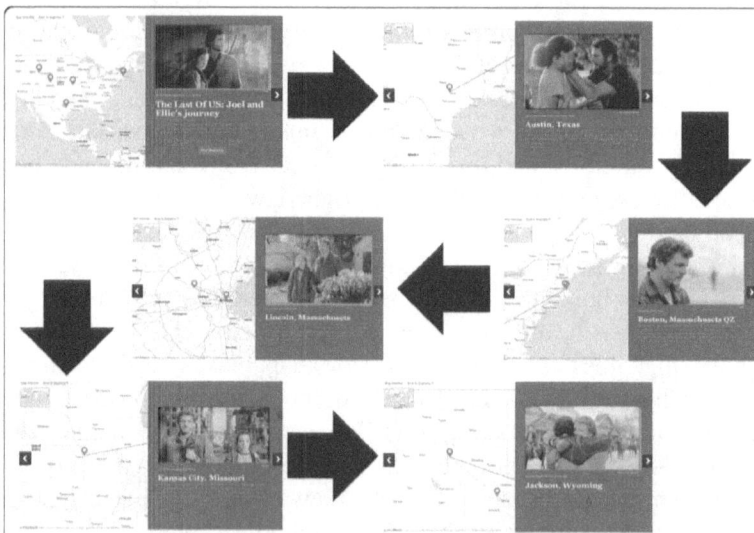

Figure 2.1 Journey in *The Last of Us*. Above depicts the interactive map that a person can navigate to take them through Ellie and Joel's journey in *The Last of Us*. The maps on the left show the location the event is happening, and on the right is a photo and caption of the event with a short description of each event. As this piece of technology involves clocking the arrow on the far-right of the screen to move from one location to the next, the bold black arrows in the picture represent what happens once those arrows are clicked. *Source*: M. Smith (2023), *The Last of Us: Joel and Ellie's Journey*, [Storymap] EDUC 550, Green State University. (All names are pseudonyms.)

with maps, therefore it was fun how the website showed me everywhere I went." Although one of the primary goals of the storymap is to depict a narrative digitally, having students become more geographically aware shows the potential versatility of using this software. Taylor also believes it is a "creative way to do a timeline project. Instead of writing a paper, or doing something boring, creating a map is a fun and engaging project to do . . . a creative and fun project teachers can give their students."

English language arts courses typically have strong writing components; by incorporating a KnightLab storymap, students can still achieve the learning goals associated with writing assignments in a way that is unique and creative. Unconventional writing assignments often raise student engagement and retention (Smagorinsky, 2018).

Students also had the opportunity to reflect on what future learners in secondary schools might struggle with when creating their own storymaps. Taylor thought that such learners would need extra assistance with "adding a location [and] coordinates," and Taylor found this aspect challenging when completing it as a part of the EDUC 550 curriculum and considered that younger students might struggle with that as well. Noting that GPS coordinates and reading maps are not consistently taught throughout PK–12, Taylor projected that educators would need to incorporate this technology to ensure they are spending adequate time preparing students to understand this feature of the storymap.

Although KnightLab makes this process fairly easy, students would need a general understanding of how GPS coordinates correlate with actual geographic locations, in addition to the logistics of plugging in a location's longitude and latitude. This would help ensure that they are logging the correct destinations, and it would further illustrate the operations of maps, both generally and digitally.

Finally, students had strong ideas about how they might use storymaps in their own future English language arts curriculum. Taylor summarized these proposals in their reflection about how they would "definitely be utilizing" KnightLab's storymap in their high school ELA classroom:

> I would have students read our assigned book, and at the end, create a storymap about it. It will show me the students' understanding of where the characters went, and what they were doing at that location.

As described, KnightLab software can be used to display any narrative with trackable locations, but when added to an existing ELA curriculum, students will be able to demonstrate in-depth knowledge of characters, storylines, authors, and history in ways that are innovative and entertaining. In a world that is becoming prominently digital, providing students with a way to illustrate knowledge in an online format will not only help to increase their English literacy, but can also be a stepping stone in establishing digital competency. As such, KnightLab's storymaps need to become a staple in secondary ELA curricula.

FilterBubbles as a Way to Link Digital Humanities to Social Media

The second assignment of the digital humanities unit asked students to analyze their personal filter bubbles. The term "filter bubble" first appeared in activist and entrepreneur Eli Pariser's book, *The Filter Bubble: What the Internet is Hiding From You*. The book was published in 2011, and in it, Pariser describes how one's internet activity (searches, location, buying, browser history, etc.) creates an algorithm that then determines what is filtered back to the user; essentially, how one uses the internet dictates their online experience. Often, these filters operate without the comprehension of the user and create complications with getting access to unbiased information. Making students aware of filter bubbles' manipulations can raise their "consciousness of the role algorithms and big data play in their lives" (McGrail, 2016, p. 25). ELA courses often have conversations about credibility, source quality, and ways to obtain reliable data; looking at filter bubbles' impact will help solidify these concepts in an applicable, relevant, and familiar domain.

In this assignment, students are introduced to the concept of filter bubbles and analyze some of the filters currently present in their social media. As they begin to understand how their actions lead to specific advertisements, news articles, and page suggestions, class conversations center around how media is personally tailored to one's interests. Students also begin to consider the ramifications of limiting information. Analyzing filter bubbles encourages students to "take control of their learning and to develop metacognitive awareness of the procedural and content knowledge required to solve problems" (McGrail, 2016, p. 22). Being able to cogently argue and thoroughly analyze are crucial skills to one's success in life and in future academic courses. After completing their analysis, students reflected on how incorporating conversations about filter bubbles could enhance secondary ELA classrooms.

A student in the course, Jackson, submitted a lesson plan titled, "*Feed:* Filter Bubbles." This lesson plan beautifully represented how one might integrate filter bubbles into an eleventh grade ELA curriculum. Spanning one hour, this student noted how a teacher would guide their class through understanding their own filter bubbles, the personal and societal implications of filter bubbles, and how, after reading *Feed* by M. T. Anderson, their overall views on this concept grew.

Jackson's lesson plan noted that, by the end of the lesson, students in the ELA class should be able to: (1) Evaluate a complex piece of literature and develop responses in connection with themselves, their world, and the novel; (2) Have conversations about the effects of technology on humanity; and (3) See multiple points of view on technology. The lesson plan noted that participating in this lesson would give students the space to personally connect with course material, gain pertinent analytical skills, and develop crucial knowledge about societal operations and standards.

Throughout the lesson, students in an ELA class will also complete various activities to help them uncover their thoughts, beliefs, and underlying assumptions about filter bubbles. Reflecting on their own filters through the video-based discussion tool, Flipgrid, students will come to class ready to discuss the questions: Do you think we are influenced by what we see today? What apps/websites did you look at? What was being filtered to you? Why do you think you are getting these specific filters? In class, these questions will be expanded upon by investigating *how* these filters impact students and society as a whole. By using the suggested discussion prompts, including: What is the significance of filter bubbles? How do filter bubbles impact privacy? How accurate do you think filter bubbles are? Jackson was able to articulate in the lesson plan how they would guide their students to think deeply about filter bubbles, including aspects of their impact, their credibility, and their authenticity. Getting students to ponder the larger "so what" questions of these algorithms will not only help them develop critical thinking skills, but will also help them think deeper and more analytically about the world's intricacies.

Jackson's lesson described how, toward the end of class, students would transition to discussing the novel, *Feed,* where they would begin to make literary connections to filter bubbles. *Feed*, a dystopian novel for young adults, centers on a society where a powerful corporation implants feeds into the public to monitor their thoughts and communications in order to manipulate the information they receive. After a short introduction to the novel and a reading of the first four chapters as a class, students would be asked to reflect on how the themes present in the text "change [their] perspectives on filter bubbles" (McGrail, 2016, p. 25). As *Feed* depicts similar themes as filter bubbles: privacy, data mining, and corporate power, comparing the narrative to existing thoughts and experiences helps solidify the content of the novel as well as provoke ideas for future application. This underscores the idea that critical ELA curriculum can connect "classroom learning to life outside [school] and to the importance of generative topics that teach for conceptual knowledge and understanding" (McGrail, 2016, p. 21). Analysis of filter bubbles assists students in developing critical understandings of their world.

Voyant as a Tool for Text Analysis

The final assignment in the digital humanities unit asked students to look at how the analysis tool, Voyant, could enhance their future secondary ELA curriculum. Voyant is a "text-analysis environment meant to support Agile Hermeneutics" (Rockwell and Sinclair, 2016, p. 8), or "talking with another person who has complimentary skills and summarizing those conversations in various ways" (ibid, p. 10). The goal of many DH softwares, including Voyant, is to create an interdisciplinary experience; by "weaving together hermeneutical things—print and electronic, text and code, interludes and reflections, narrative and interaction" (ibid, p. 3), DH adds new depth, perspectives, and connections to projects. Since Voyant is fairly complex, the assignment in EUC 550: Technology for Teachers began with conversations around what it means to read, to analyze, and to ask questions. Because DH produces quantitative data (versus the qualitative data that the humanities are used to associating with its research), it was important to discuss what kinds of questions students should be asking to obtain statistical data about a piece of writing.

After obtaining a general understanding of how to perform an analysis in Voyant, students started working in the actual program. Although Voyant has numerous tools, the class focused on the five that would be the most applicable to the analysis of ELA curricular materials: Vocabulary Density/Average Words Per Sentence, Cirrus Tool, Trends Tool, Contexts Tool, and Correlations Tool. Together, these tools examine different features of texts, including sentence structure, word choice/repetition, organization, theme, and general patterns. Students were asked to analyze a text with at least three different tools and then reflect on their experience.

One student, Taylor, chose to select the novel, *Crime and Punishment*, by Fyodor Dostoyevsky, to see if they could understand the basic storyline only from Voyant Tools (see figure 2.2). Taylor describes their Voyant experience:

> When looking at the text, I used the [Cirrus Tool] to see which words were used most frequently: Raskolnikov, Razumihin, Dounia, Petrovich, man, suddenly, don't, come, and know. From this, I already know the names of the major characters.
>
> Next, I used the Document Terms function and found that mentions of Dounia peak in the middle of the story and then drop off, implying she was involved in a major plot point and then faded into relative obscurity one way or another.
>
> Finally, by using the Phrases tool, I found the following recurring phrases: "in the police office". "government quarters", "he was drunk", "I am a murderer", "his mother and sister", "the commisseriat". From these tools, I've learned the names of the main characters (especially that Raskolnikov is the protagonist),

Figure 2.2 Cirrus Tool Word Cloud. The Cirrus Tool creates a word cloud of the most used words in the text being analyzed. The larger the words appear, the more frequently they are used in the text. This tool can be edited to exclude unwanted words, also known as stop words, to make a run more tailored to one's goals. *Source*: T. Ellington (2023), Cirrus tool word cloud, EDUC 550: Green State University. (All names are pseudonyms.)

that the events of the novel revolve around the interaction between Raskolnikov (and possibly his family) and a Russian government entity. Through the mention of drunkenness and murder, it may suggest that the protagonist himself has committed or been involved in a crime for which he's being pursued by the authorities.

By analyzing this novel using tools in Voyant, Taylor shows that Voyant can be used to gain an understanding of major plot points. In addition, students can become familiar with themes, character relationships, major settings, and how these interweave. Because Voyant illustrates unique connections, such as how characters might relate to certain themes, it offers a unique perspective on understanding literature. Instead of backing claims with conceptual understandings of narratives, students can unlock deeper and more abstract knowledge about course material (see figure 2.3).

Another student, Karsyn, decided to run an eleventh-grade research paper on Doris Lessing's "To Room Nineteen" to see if they proved their thesis throughout their essay. Karsyn reflects on their Voyant run:

My original thesis for the paper was: Through the use of colorful imagery and the disturbing symbolism of Susan Rowlings' suicide, Lessing's "To Room Nineteen" exemplifies the constricting and damaging gender roles of the 1960s, showing that forcing women into such a confining role can take a toll on their

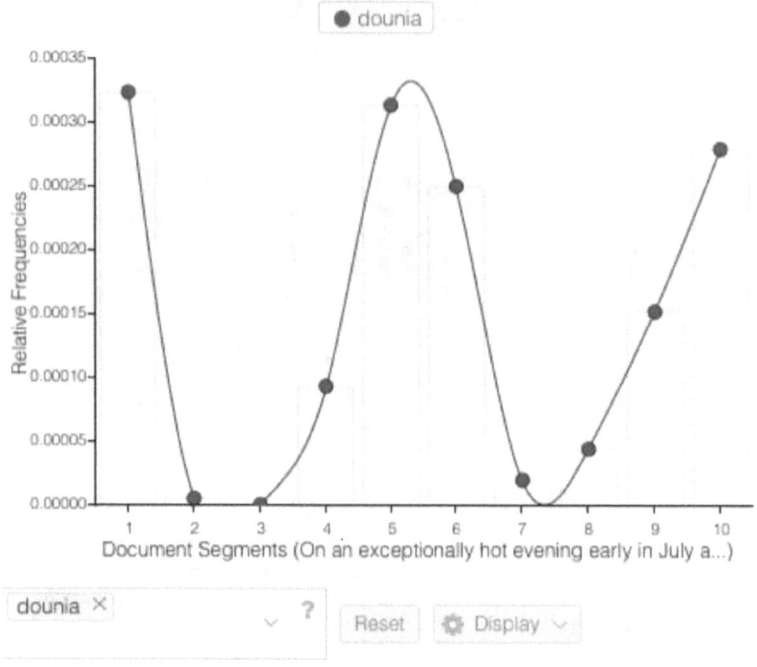

Figure 2.3 Word Frequency. The trends tool allows users to look at the frequency of words throughout a text. The user can look up specific word/s in the search bar at the bottom left corner, which can allow for clarity on the word/s usage in the passage. *Source*: T. Ellington (2023), Cirrus tool word cloud, EDUC 550: Green State University. (All names are pseudonyms.)

mental health. I used the Cirrus tool first to see what words I used the most in the paper. I did this to see if my paper did centralized around my original thesis statement. I then picked 5 of the most used words that fit my thesis and plugged them into the trends and context tools. The words I chose were women, roles, suicide, mental, and freedom. With the trends tool I was able to see where I used these words the most often in my paper and it helped me see what sections did a good job of supporting my thesis and which sections I might have needed to look back on to add to. When I looked at the context tool I was able to pick out and see where I used the words I chose to focus on and how I used them in my paper. This gave me a better idea of what my focus was on while I was writing the paper and helped me to pick out when the words were effectively used to support my thesis and when the words could have been used better to support my thesis.

Karsyn's Voyant run depicts how this tool can aid in the editing process. Many students struggle with understanding how to dissect an essay for changes, usually resulting in more micro-level edits, like grammatical, versus macro-level edits, like structural or conceptual. Using Voyant as an editing tool could provide clarity on how well central arguments are supported, revealing visually where specific parts of the essay deviate from the primary claim. As such, by implementing Voyant into ELA revision practices, students have the potential to broaden both their abilities and appreciation of the editing process, which in turn should make them more diligent writers moving forward.

Along with reflecting on their Voyant experiences, students also considered how they pictured the program being used in the secondary ELA curriculum. Taylor believes that it would be the "most useful for post-production analysis of student/peer work." Giving students a "thorough way to understand the weaknesses of their papers, such as when they overuse certain words or phrases could help make the peer review process smoother and more appealing." Karsyn also pictured Voyant being used in a peer review process, as they note students can "double check if their paper is staying on topic and where they might need to refocus their paper . . . and if they become repetitive." Both Taylor and Karsyn validate Voyant's use as a peer review and editing tool through their Voyant-run analyses and reflections. Incorporating this piece of DH in ELA peer review practices will give students an innovative, yet thorough, way to learn effective editing skills, which can then be applied to all future writing.

MOVING FORWARD: DIGITAL HUMANITIES AND STREAMS

As the use of KnightLab, Filter Bubbles, and Voyant illustrates, incorporating DH into pre-service teachers' understandings of the curriculum goes beyond, as EDUC 550 is titled, technology for teachers. Technology in and of itself is taught and used in the course, but through the use of digital humanities, pre-service teachers are encouraged to see technology as something generative rather than just a tool or an end.

Digital humanities encourages students to develop skills that are embedded in the discipline of ELA, but do this in new ways. KnightLab, for example, encourages story mapping and writing in unconventional ways, yet still draws on the elements of narrative, including plot, characters, and storylines. Analysis of FilterBubbles fosters pre-service teachers' understanding that ELA is not just about teaching the skills of reading and writing, but also about helping students develop a critical understanding of their world. Asking students to question the media and messages they consume has the potential to raise students' awareness of their world. Voyant, offering a different kind of text analysis, highlights the components of text in new ways, prompting students to ask questions about what comprises a text. Digital humanities enhance the goals of a rich ELA curriculum and learning experiences.

Through the application of digital humanities, we seek to transform the higher education classroom through equitable practices. Depicting change through the use of DH brings attention to fostering a more inclusive and equitable learning environment. Seeking to create an understanding of how we can more fully humanize our students, the use of DH can generate connectedness and belonging, and can inspire change in higher education.

The integration of digital humanities within the discipline of English language arts does not just add a digital dimension; rather, it changes the way we go about inquiring in the English language arts. By incorporating DH, students can achieve the Common Core Standard of close reading. This standard states that students in grades K-12 need to be able to: "read closely to determine what the text says explicitly and make logical inferences from it; cite specific textual evidence when writing or speaking to support conclusions drawn from the text" (Hancher, 2016, p. 126). Students in English language arts classrooms were often asked to consider how the texts they encountered impacted them personally; but with the focus on close reading, "the goal is to understand what the author is doing and accomplishing and what it means ... it is not to respond personally to what the reader is doing ... tasks should require careful scrutiny of the text and specific references to evidence from *the text itself* to support responses" (Hancher, 2016, p. 126).

The support of DH in close reading within the context of an ELA classroom allows students to ask deeper questions, observe more abstract connections, strengthen their writing practices, and become better prepared for the current expectations of college curriculum. Students will see the connections texts have to themselves while adding in the value that texts bring on their own and to society as a whole. DH brings new "abilit[ies] to access and organize minute particulars [which] can do more than inform our understanding of an isolated text ... it can generate new knowledge, not merely sustain old prejudices" (Hancher, 2016, p. 128). These innovative curricular pathways will be groundbreaking for any student who wants applicability, relatability, stimulation, and rigor within the curriculum with which they engage.

From its inception, DH has been encompassed by conversations of humanities pedagogy; however, there are numerous benefits to expanding these dialogues to include both math and science. Adding DH to math and science curricula can foster engagement, self-esteem, and the overall practicality of these subjects. A study by Lin-Siegler et al. (2023) surveyed 300 high school STEM students (150 math; 150 science) to help discover what impacted their motivation and performance in these classes. Results showed the most common theme "involved the learning process itself; more than half of the events in the narratives were about not understanding the material, insufficient motivation or effort, not employing effective study or test preparation strategies, or experiencing anxiety about tests or about performance more generally" (Lin-Siegler et al., 2023, p. 114). DH could be an excellent tool to enhance comprehension and increase student motivation. Because DH often takes a design-based learning structure, it automatically adds more structure and applicability to assignments and curriculum. Using DH tools that utilize topics students are already familiar with has the potential to help create projects that are more personalized, and tailored to individual interests.

In addition, these design-based projects will allow students to practice and learn the material in a more engaging way than through just rote studying. Finally, by "asking scholars to describe their pedagogical methodologies within an interdisciplinary space," students can better connect their math and science pedagogy with the larger scope of academia as a whole (Clement, 2012, p. 377). This is often difficult for students to do, so giving them a creative space to explore these tough concepts in a way that shows their applicability could enhance their overall motivation and willingness to learn these subjects. A goal of DH is to bring subjects together; it is interdisciplinary. As such, because "different contexts require different pedagogies [. . .] digital humanities is a meeting house for these many minds" (Clement, 2012, p. 380). Therefore, adding DH to math and science curricula would only benefit students.

Today, it is more important than ever that ethical uses of technology are embedded within the ELA curriculum. Each year, society's dependence on technology increases, and with that comes an expectation of students' use of high-level digital literacy. Integrating digital humanities into ELA classrooms will not only give students a foundation for the use of technology but expose them to thinking critically about their technology use. By thinking through their intentions, connections, and the overall impact of their DH work, students will become more intentional with their technology use. DH also reveals abstract connections that would otherwise remain unknown, and by using DH software, the curriculum will increase in relevance not only to the world around them but also to the other subjects they are required to take. Instead of having segregated subjects and topics, DH can interweave digital literacy, intentionality, and interdisciplinarity, all of which are crucial to developing a deeper understanding of the intricacies of the world and the role academics can have in the world.

REFERENCES

Anderson, M. T. (2022). *Feed.* Cambridge, MA: Candlewick Press.

Clement, T. (2012). Multiliteracies in the undergraduate digital humanities curriculum: Skills, principles, and habits of mind. In B. D. Hirsch (Ed.), *Digital humanities pedagogy: Practices, principles and politics* (1st ed., Vol. 3, pp. 365–388). Open Book Publishers. http://www.jstor.org/stable/j.ctt5vjtt3.20

Hancher, M. (2016). Re: search and close reading. Digital humanities at community colleges. In M. K. Gold & L. F. Klein (Eds.), *Debates in the Digital Humanities 2016* (pp. 118–138). University of Minnesota Press.

Jones, S. E. (2016). The emergence of the digital humanities (as the network is everything). In M. K. Gold & L. F. Klein (Eds.), *Debates in the Digital Humanities 2016* (pp. 16–31). University of Minnesota Press.

Lin-Siegler, X., Lovett, B. J., Du, Y., Yamane, K., Wang, K., & Hadis, S. (2023). What experiences constitute failures? High school students' reflections on their struggles in STEM classes. *Annals of the New York Academy of Sciences.*

Northwestern University, (2023). *Knight lab.* Knightlab, https://knightlab.northwestern.edu

McGrail, A.B. (2016). The "whole game:" Digital humanities at community colleges. In M.K. Gold & L. F. Klein (Eds). *Debates in the Digital Humanities 2016* (pp. 16–31). University of Minnesota Press.

Rockwell, G. and Sinclair, S. (2016). *Hermeneutica: Computer-assisted interpretation in the humanities.* MIT Press.

Smagorinsky, P. (2018). *Teaching English by Design: How to create and carry out instructional units.* Heinemann.

Chapter 3

Computer Science Curriculum for Culturally and Linguistically Diverse Students

Clare Baek, Sharin Jacob, Dana Saito-Stehberger, Leiny Yesenia Garcia, Santiago Ojeda-Ramirez, and Mark Warschauer

STREAMS: INTEGRATING LANGUAGE AND LITERACY TO STEM

The integration of language and literacy into STEM subjects is an effective strategy to create an inclusive learning environment that supports the unique challenges experienced by diverse learners. By infusing STEM curricula with reading, writing, speaking, and disciplinary literacy practices, educators can foster inclusive learning environments that promote learning outcomes for all students, particularly those from culturally and linguistically diverse backgrounds. Finding ways to develop and implement STEM curricula is crucial, as students from culturally and linguistically diverse groups often struggle in STEM courses and are underrepresented in STEM fields. For example, Latinx students are underrepresented in STEM fields due to reasons including having unequal learning opportunities in K–12 education (Gautreau et al., 2019; Owens & Ramsay-Jordan, 2021). Further, although the number of students speaking English as a second language has increased exponentially in recent years, there has been consistent evidence that these students achieve lower academic learning outcomes compared to their native English-speaking peers (LaCosse et al., 2020; Maarouf, 2019).

    STEM subjects incorporate technical terminologies and complex linguistic texts, which can pose barriers to English language learners, others with language difficulties, and those who may have had less exposure to learning in STEM contexts (Maarouf, 2019). Successful STEM learning requires an understanding of discipline-specific vocabulary and analysis of complex texts and discourse (Kamberelis et al., 2014; Lefever-Davis & Pearman, 2015). However, traditional STEM instruction's reliance on abstract concepts, communicated through technical vocabulary in complex texts, exacerbates the difficulties faced by linguistically diverse students (Israel et al., 2013). To support the needs of diverse students and foster their success, it is critical to integrate language and literacy components into STEM curricula (Feng et al., 2020; Rivera & Molinda, 2017). Integrating language and literacy components into STEM curricula engenders a cyclical relationship such that it would support students' understanding of STEM concepts and students' STEM learning experiences would promote their language and literacy skills at the same time (Israel, 2013).

INTEGRATING LANGUAGE AND LITERACY WITH COMPUTATIONAL THINKING

In this digital age, proficiency in coding is integral for academic success and civic engagement. Integrating language and literacy with coding can enhance students' computational thinking skills and foster their ability to communicate, collaborate, and solve real-world problems effectively, thus preparing them for success in both academic and professional spheres.

While coding is conventionally perceived as a technical process for instructing computers, it serves as a dynamic medium for students to express their creativity (Burke & Kafai, 2010; Lowe & Brophy, 2019; Smith et al., 2020). This process inherently requires computational thinking skills, which are a set of cognitive processes involving algorithmic and systematic thinking to formulate questions and solve problems (Jacob & Warschauer, 2018; Song et al., 2021; Weber, 2018). Computational thinking applies to problem-solving in various contexts and extends beyond STEM fields.

Brennan and Resnick's (2012) computational thinking framework delineates computational thinking into three dimensions: computational concepts, computational practices, and computational perspectives. Computational concepts describe the ideas students engage with when creating a program. For example, computational concepts include writing sequential instructions for a computer to execute intended behaviors (Schute et al., 2017). Computational practices describe the habits students develop as they engage with computational concepts, such as remixing existing products to create new ones (Brennan & Resnick, 2012). Computational practices encompass being incremental and iterative, testing and debugging, reusing and remixing, and embracing abstraction and modularity (Brennan & Resnick, 2012). Incremental and iterative represent the processes of designing and implementing solutions. Testing and debugging pertain to the problem-solving process when the code does not work as intended. Reusing and remixing involve drawing ideas from existing projects and modifying them with one's own. Abstraction and modularity describe assembling smaller parts to create a more complex whole. Computational thinking perspectives refer to the viewpoints students form about the world and themselves, such as their ability to express ideas by creating a product, collaborating with others to share ideas, and applying them to address and solve real-world problems (Brennan & Resnick, 2017; Schute et al., 2017).

Thus, computational thinking can be harnessed to foster language and literacy skills, enabling the creation, refinement, and presentation of narratives (Jacob et al., 2018; Schute et al., 2017). Jacob and Warschauer (2018) present a three-dimensional framework: computational thinking as literacy defines computational thinking as literacy itself; computational thinking through literacy examines how to leverage students' existing literacy skills to develop their computational thinking skills, and conversely; literacy through computational thinking investigates how to mobilize students' computational thinking skills to develop their literacy skills.

However, students from diverse cultural and linguistic backgrounds continue to be underrepresented in the realm of computing. Latinx students, in particular, face barriers such as a lack of Latinx role models in CS (Computer Science), limited CS course offerings in Latinx-neighborhood schools, and pedagogies that do not align with their cultural values (Denner & Campe, 2023; Jacob et al., 2020). Culturally relevant CS curricula can empower Latinx students by leveraging their community knowledge and cultural identity (Luna & Martinez, 2013).

A CULTURALLY RELEVANT CS CURRICULUM INTEGRATED WITH LANGUAGE AND LITERACY

A CS curriculum offers a unique platform for integrating culturally responsive pedagogy, as it can provide visual and interactive content that engages students' creativity. A culturally relevant CS curriculum involves connecting coding concepts to students' experiences, drawing from their cultural heritage and contemporary youth culture, and offering opportunities for students to create projects related to their own lives (Franklin et al., 2020). Culturally relevant pedagogy plays a pivotal role in creating inclusive learning environments that honor students' cultural backgrounds and experiences. By integrating cultural elements into teaching, this pedagogy makes learning more meaningful and relevant for ethnically and culturally diverse students (Brown-Jeffy & Cooper, 2011). A culturally relevant CS curriculum addresses not only the need for technical skill development but also promotes the inclusion of diverse cultural narratives, enhancing student engagement and comprehension.

Guided by culturally relevant pedagogy, the CS curriculum presented in this chapter seeks to nurture diverse elementary school students' creativity, computational thinking, and literacy through projects that hold personal significance. Particularly, the curriculum was designed to support school districts with a large percentage of Latinx and multilingual students. The curriculum is taught in Scratch, a block-based programming language with interactive visual tools and multimedia features. Scratch is an ideal programming environment for younger students, as they can easily drag and drop blocks to create a wide range of projects while learning essential programming concepts such as sequences, loops, events, and conditionals (Brennan & Resnick, 2012). The lessons throughout the curriculum culminate in two projects about the students themselves. In the first project, "Name Poem," students craft a poem by selecting an adjective for each letter of their name, each representing a personal trait. In the second project, "About Me," students describe themselves using various Scratch blocks and Sprites (characters in Scratch). In both projects, students not only acquire CS skills but also engage in computational thinking by narrating their personal experiences and expressing their creativity. Further, there are role model lessons about computer scientists from culturally diverse backgrounds, including Latinx backgrounds. The role model lessons include videos of the Latinx computer scientists discussing how they overcame challenges (e.g., challenges experienced as a person of color or a woman in the field) to pursue their academic and career trajectories. These role model lessons are crucial for Latinx students as the underrepresentation of Latinx students in CS can be attributed to factors including a lack of Latinx role models in the field (Jacob et al., 2020).

Central to culturally relevant pedagogy is the acknowledgment of students' backgrounds as assets rather than barriers to learning. By embedding language and literacy practices within STEM education, educators can create a learning environment that resonates with students' lived experiences (Casler-Failing et al., 2021). Learning environments that are relevant to students' lives can increase engagement and elicit positive attitudes among culturally and linguistically diverse students, as well as increase their collaboration with peers (Rivera & Molinda, 2017). The process of writing code to create projects that describe their names or present stories about themselves simultaneously involves computational thinking, language, and literacy skills. The "Name Poem" and "About Me" projects encourage students to activate adjectives they already know or acquire new ones as they construct narratives about themselves. A list of adjectives that students can use is provided in the student workbook. These projects encourage students to interact with vocabulary in a meaningful context, which is a strength of a STEM curriculum integrated with literacy (Wood et al., 2011). Working on these projects naturally involves students sharing their interests and expressing themselves, which creates an authentic collaborative learning environment while promoting communication skills (Israel et al., 2013).

In this curriculum, students promote their literacy skills while engaging in coding activities. In debugging activities, students practice writing an algorithm step-by-step to draw the same house as the given drawing. This activity not only introduces the concept of debugging to students but also promotes their literacy skills by requiring them to write clear instructions in order, using the correct words. Also, as students are naturally engaged in problem-solving through this debugging activity, it can encourage their motivation and positive attitudes in addition to acquiring technical skills (Lefever-Davis & Pearman, 2015). Further, sentence frames are incorporated into the student workbook to help students construct meaningful sentences, thereby enhancing their language skills. For example, the workbook provides sentence frames with blanks for students to describe their coding process, such as "I used (name of the blocks) blocks, which caused (name of the action) to happen."

CASE STUDY OF THE INTEGRATED CS AND LITERACY CURRICULUM

We describe a case study of the implementation of the integrated CS curriculum we outlined above. The participants are fourth-grade students from a large school district in Southern California, with approximately 95 percent Latinx students and about 50 percent designated as English learners. With this case study, we explore how Latinx students engage in linguistic practices and computational thinking through a culturally relevant CS curriculum. We wanted to examine students' coding and literacy learning processes, the effectiveness of the curriculum in improving literacy and computational thinking skills, and the unique strengths that Latinx students bring to their learning.

Semi-structured interviews were conducted with 12 students who had undergone one year of CS curriculum instruction. The students interviewed were all Latinx and designated as English language learners. The topics for the questions asked during the interview were about how the projects ("Name Poem," "About Me") describe who the students are, why and how they chose certain characters, why and how they chose to use certain blocks, and their trial-and-error process as they created the projects. During the interview, the students shared the projects they made using Scratch with the interviewers as they answered the questions. The interviews averaged 25–30 minutes in duration.

We transcribed the audio recording of each interview verbatim using transcription software. Thematic analysis of interview transcripts was conducted following an inductive approach (Braun & Clarke, 2006). Initial codes were generated iteratively, and relevant excerpts were grouped under each code. These codes were then organized into themes, reviewed collaboratively by the research team, and refined iteratively. Excerpts were reexamined to identify additional codes aligning with the established themes.

Four overarching themes emerged from the interviews. As students created projects that were relevant to their everyday lives and, therefore, meaningful, they developed and refined their linguistic and computational thinking skills. We describe the details of each theme below with quotes from student interviews. The verbatim quotes presented below are selected excerpts from the interviews.

Connecting the Coding Process to Solving Real-Life Problems

Students discussed how they could use coding to solve real-life problems and make a positive contribution to helping others. For example, one student expressed a desire to create a game in Scratch to "teach" people about "unexpected things in the forest, such as nature or animals." Another student related the process of finding a lost item to assembling blocks of code, step by step, to figure out where the item might be located. This student saw a connection between following a step-by-step sequence of writing code and solving real-life problems.

Students also connected the perseverance and emotional regulation they employed during the debugging process in Scratch to problem-solving in real life. They indicated that when things didn't work out as intended while coding, they should keep trying. Moreover, they found value in their parents' advice about the problem-solving process, which contributed to their persistence as they engaged in debugging.

> I feel like, um, I should not get upset. I should just keep trying and trying and trying and, because you learn from your mistakes. My mom always told me that and my dad.

Expressing Ideas and Visions through Computational Thinking Practices

Students effectively utilized computational thinking concepts to convey their ideas and narratives while actively working on projects in Scratch.

Their engagement with the coding process naturally facilitated the practice of computational thinking skills. To articulate their ideas and stories in alignment with their creative vision, students had to navigate the process of configuring their Sprites' appearance and actions, encompassing aspects such as size, speed, location, color, and sequence. In this way, students actively participated in computational thinking practices, essentially translating their personal preferences and interests into their Scratch projects.

As students created projects that represented their names and themselves, they carried out an interactive process, fine-tuning the size of the letters to match their creative vision. While manipulating the letters in their "Name Poem" projects to attain their desired placements, they crafted iterative instructions. Throughout this process, students actively embraced trial and error, experimenting with instructions, closely observing how the letters responded, and making the necessary adjustments to achieve their intended outcomes.

> There was a bug in mine because I had put these (letters) too big. I had them in 95 and I, and I, there was no space for P and H, so, I, made them smaller. I made them to 90 and I put them more like closer, a little bit more closer.

One student described his trial-and-error steps:

> I just started over and try to take step what I did wrong and tried to look and then I found out what I did wrong, and so I put the number too much to go to the other side.

Another student performed mathematical calculations to manipulate the letters' location when they were clicked.

> [wanted to have the letter N move] like in the front and then go back. Okay. So I added move 50 steps, and then put move minus 50 steps again. So when it goes in the front, and then when I press it again it stops and it goes back.

Students employed an incremental and iterative approach to position the letters of their names precisely, achieving the desired appearance (e.g., glide versus jump) and ensuring that the letters were synchronized in the preferred order.

> First I want to . . . use the glide block, so then I could show I pushed it and put it up. I want to show for the x and y . . . so when I clicked it, it was going up. First it was gonna go up, then say, click on the down arrow to put me down. And . . . when it was going down, I wanted to say I picked a soccer ball.

Students composed instructions in Scratch to specify the actions they wanted the computer to execute, including determining what Sprites should say and the duration for which they should deliver the text. For instance, one student explained his interaction with the computer, stating, "I examined this one and instructed it to say 'Hello' for two seconds. Then, I placed it here . . . "

Expressing Identities and Interests through Computational Thinking Practices

Students expressed what they liked and who they are by making projects about themselves in Scratch. By crafting projects that reflected their identities and interests, students honed their coding skills while making personal and meaningful connections to their work.

> We searched up a picture that we wanted for our Scratch . . . you look for the picture that you wanted and then you click it and that's how you put the one, the picture that you wanted.

One student mentioned that, while working on his "Name Poem," his teacher encouraged him to place the letters of his name "anywhere you want" on the Scratch page. This opportunity for creative expression added a personal and meaningful dimension to his learning.

> Because whenever you click the Sprite, it will talk about more like me telling what I like to do, something, things that I like.

Students enjoyed the process of making things that are about them and that represent their identities. For example, one student created a project where a letter in his name changes to his favorite color when clicked. Students described their favorite part of coding in Scratch as the ability to create something that resonated with their individuality.

> My favorite part is like putting, finding a picture, and like, putting letters, what your name says about you.

One student shared her experience of creating a project centered around a soccer ball because she enjoys playing soccer with her siblings. In her project, she represented a soccer ball bouncing up and down, mirroring the activity she loves. She accomplished this by coding the action to be triggered when the ball is pressed with a keyboard key and also programmed the ball to move up and down using the motion blocks in Scratch.

> For the soccer ball . . . I put that and I could play with my siblings, but first I touched, I pressed on the ball and went up. Then I asked to push the down

arrow, it pushed the down arrow. And then it says that I liked to play with my siblings outside.

Students remixed existing Scratch projects to express their own ideas and creativity. They began with others' projects for inspiration and then customized them by changing features like colors and pictures to make the projects their own.

> I saw this project that wasn't mine . . . so I took off all the letters . . . that's what I put my letters on it. And, and I looked for the one, the purple, um, block that said said something and then for, okay how many seconds? And I put that in, I choose my own, my own picture.

Promoting Literacy through Expression and Computational Thinking

While creating projects about themselves, students searched for adjectives that described themselves along with their favorite things. For instance, one student chose "joyful" to represent the letter "J" in their name. Another student included "A" for artist and "R" for running to depict their interests. Students utilized the adjective list in the curriculum's workbook to find new words to describe themselves. They engaged with these adjectives by saying the words out loud to peers and teachers when describing their projects, and spelling them correctly as they coded the words in their Scratch projects.

As students worked on making specific actions happen in Scratch, they also applied their knowledge of synonyms. For instance, when one student needed to hide their characters, they initially browsed the "motion" blocks but later realized that the relevant block was in the "looks" section, saying, "I thought it was in the looks section, and one of these blocks said 'Hide.'" Another student wanted their characters to "disappear," so they specifically looked for the "Hide" block.

Students also learned discipline-specific vocabulary through the lessons and naturally engaged with them as they created projects. For example, students learned computational thinking-related words including algorithm, program, sequence, and debugging. Students also learned technical coding words such as "loops" and "event." As students created their projects in Scratch, they were exposed to these words repeatedly. For example, students asked their peers or teacher for help when their code did not work correctly and naturally used technical words to describe what they intended to create and what was not working.

DISCUSSION

The culturally relevant pedagogy empowered the students in this study, a community that has traditionally been underrepresented in CS education. Through culturally relevant lessons that connected with their lives, students found the agency to learn necessary skills and persistently work on projects that mattered to them. These lessons also cultivated a critical awareness that coding skills could be applied to solve real-life problems and contribute to the world (Ladson-Billings, 1995; Tissenbaum et al., 2019). Students began to see coding as a means of engaging with the world around them and applied their coding experiences to solve real-life problems. Furthermore, students naturally enhanced their language and literacy skills as they practiced computational thinking skills.

To create projects that represented meaningful aspects of their daily lives, students engaged in computational thinking practices and utilized various computational thinking concepts. Students figured out the appropriate parameters for their Sprites through an iterative process of experimenting with different parameters and making adjustments until they achieved the desired results. When a project did not run as intended, students conducted a trial-and-error process to debug errors. Additionally, students remixed others' projects to gain initial ideas and customized them by modifying Sprites, blocks, and parameters. Through a culturally relevant curriculum, students did not feel that they were being required to learn computational thinking concepts; instead, they felt they were creating what they wanted in ways they preferred. Thus, students did not just learn to code; they coded to create something meaningful to them, highlighting how culturally relevant pedagogy motivates students to develop computational thinking concepts (Aronson & Laughter, 2016).

The role of family in students' motivation and perseverance in coding was evident. Students practiced their computational thinking skills by drawing upon the cultural value of *familisimo*. For Latinx students, family values play a pivotal role in their academic motivation and outcomes (Azpeitia & Bacio, 2022; Esparza & Sánchez, 2008). When encountering difficulties during coding, students recalled what their parents had told them about not giving up and continuing to problem-solve and debug. Their parents' encouragement instilled in them the value of perseverance. Additionally, when creating projects about themselves, students depicted activities they engage in with their family members through coding.

Students' computational thinking practices were inherently intertwined with their linguistic practices. When students wrote a sequence of instructions to a computer, they gained an understanding of sentence structure and syntax, paralleling language learning practices (Bers et al., 2014; Jacob &

Warschauer, 2018). As students described their coding processes, they demonstrated sequential steps in writing instructions to a computer while ensuring the proper semantics and syntax structure. Students had to plan the sequence of actions they wanted Sprites to perform, select the appropriate blocks for each step, and arrange them in the right order to achieve their intended actions. Furthermore, students leveraged their existing literacy skills to develop computational thinking skills, such as recognizing synonyms to find the correct blocks and arranging them in the correct sequence. The process of students searching for the correct Scratch blocks in the interface activated their vocabulary. The activation of vocabulary, where learners access and use the words they have learned, is crucial in second language acquisition (Gu, 2003; Teng, 2022).

Our findings present an effective integration of language and literacy into STREAMS, particularly through a culturally relevant CS curriculum. For example, the infusion of literacy into STEM subjects inherently supports the "Reading" component by incorporating literacy practices critical to understanding scientific and technological concepts. The intersection of technology, engineering, and math with literacy activities, as demonstrated in the "Name Poem" and "About Me" projects, allows students to develop computational thinking alongside language skills, which shows a blend of STREAMS. In these two projects, students promoted their math skills as well by figuring out numbers and sizes to move and position their Sprites in the program. Also, the coding process involved engineering as students tested their code to ensure it worked properly, used mathematical and computational representations to problem-solve, and learned to communicate clearly about their ideas to others during the debugging process (Bybee, 2011). Moreover, the inclusion of art and creativity within the CS curriculum—especially through the use of Scratch to express personal and cultural narratives—resonates with the "art" in STREAMS, highlighting that computational thinking does involve creativity. This integration is particularly empowering for students from diverse backgrounds as they could see themselves as innovators creating personally meaningful projects within the STREAM fields.

The social sciences component of STREAMS is addressed in our findings through the cultural relevance of the curriculum. By valuing students' backgrounds and incorporating these into the CS curriculum, educators acknowledge the importance of understanding cultural and social influences on learning. Indeed, to create an effective curriculum to integrate STEM, it is essential for educators to know the students, including what is culturally important to students and what is individually appropriate for students (Rivera & Molina, 2017). The curriculum that allows for students to share their cultural backgrounds would enable teachers to build knowledge of students' backgrounds, which would help them tailor their pedagogical practices accordingly (Kamberelis et al., 2014).

CONCLUSION

The significance of this work lies in its ability to transform STEM learning by foregrounding the need for cultural relevance and linguistic inclusivity. STREAMS education, therefore, becomes not only multidisciplinary but also multicultural, allowing students and teachers to build a more holistic engagement with the world while developing diverse perspectives. The integration of language and literacy into a culturally relevant CS curriculum enhances the STREAMS framework by showing that every discipline of STREAMS is not only interconnected but also interdependent with cultural and linguistic components. This multidisciplinary and multicultural approach not only promotes diverse students' STEM and literacy outcomes but also prepares all students to be engaged citizens and creative thinkers who can understand and value different cultures and perspectives.

Acknowledgment

This work was funded by the National Science Foundation under Grant Number 1923136.

REFERENCES

Aronson, B., & Laughter, J. (2016). The theory and practice of culturally relevant education: A synthesis of research across content areas. *Review of Educational Research*, 86(1), 163–206.

Azpeitia, J., & Bacio, G. A. (2022). 'Dedicado a Mi Familia': The role of familismo on academic outcomes among Latinx College students. *Emerging Adulthood*, 10(4), 923–937.

Bers, M. U., Flannery, L., Kazakoff, E. R., & Sullivan, A. (2014). Computational thinking and tinkering: Exploration of an early childhood robotics curriculum. *Computers & Education*, 72(March), 145–157.

Braun, V., & Clarke, V. (2006). Using thematic analysis in psychology. *Qualitative Research in Psychology*, 3(2), 77–101.

Burke, Q., & Kafai, Y. B. (2010). Programming & storytelling: Opportunities for learning about coding & composition. In *Proceedings of the 9th International Conference on Interaction Design and Children*, 348–351. IDC '10. Association for Computing Machinery, New York, NY.

Brennan, K., & Resnick, M. (2012). *New frameworks for studying and assessing the development of computational thinking*. Of the 2012 Annual Meeting of American Education Research Association.

Brown-Jeffy, S., & Cooper, J. E. (2011). Toward a Conceptual Framework of Culturally Relevant Pedagogy: An Overview of the Conceptual and Theoretical Literature. *Teacher Education Quarterly* 38 (1), 65–84.

Bybee, R. W. (2011). Scientific and engineering practices in K–12 classrooms: Understanding a framework for K–12 science education. *Science and Children*, 49(4), 10.

Casler-Failing, S. L., Stevenson, A. D., & King Miller, B. A. (2021). Integrating mathematics, science, and literacy into a culturally responsive STEM after-school program. *Current Issues in Middle Level Education*, 26(1), 3.

Crane, P. R., Talley, A. E., and Piña-Watson, B. (2022). This is what a scientist looks like: Increasing Hispanic/Latina women's identification with STEM using relatable role models. *Journal of Latinx Psychology*, 10(2), 112–127.

Denner, J., & Campe, S. (2023). Equity and inclusion in computer science education: Research on challenges and opportunities. In S. Sentance, E. Barendsen, N. R. Howard, & C. Schulte (Eds.), *Computer science education: Perspectives on teaching and learning in school* (p. 85).Bloomsbury Publishing.

Esparza, P., & Sánchez, B. (2008). The role of attitudinal familism in academic outcomes: A study of Urban, Latino High School seniors. *Cultural Diversity & Ethnic Minority Psychology*, 14(3), 193–200.

Feng, S., Garimella, U., & Pinchback, C. (2020). Integrating literacy into STEM education: Changing teachers' dispositions and classroom practice. *Journal of STEM Teacher Education*, 55(1), 3.

Franklin, D., et al. (2020). Scratch encore: The design and pilot of a culturally-relevant intermediate scratch curriculum. In *Proceedings of the 51st ACM Technical Symposium on Computer Science Education* (pp. 794–800). SIGCSE '20. Association for Computing Machinery, New York, NY.

Gautreau, C., Brye, M. V., Mitra, S., & Winstead, L. (2022). Engaging Latinx students in STEM learning through chemistry concepts. *Journal of Latinos and Education*, 21(5), 482–489.

Gu, P. Y. (2003). Vocabulary learning in a second language: Person, task, context and strategies. *Tesl-Ej*, 7(2), 1–25.

Israel, M., Maynard, K., & Williamson, P. (2013). Promoting literacy-embedded, authentic STEM instruction for students with disabilities and other struggling learners. *Teaching Exceptional Children*, 45(4), 18–25.

Jacob, S., Nguyen, H., Garcia, L., Richardson, D., & Warschauer, M. (2020). *Teaching computational thinking to multilingual students through inquiry-based learning* (Vol. 1). IEEE.

Jacob, S., Nguyen, H., Tofe-Grehl, C., Richardson, D., & Warscahuer, M. (2018). Teaching computational thinking to English learners. *NYS TESOL*, 5(2), 12–24.

Jacob, S., Garcia, L., & Warschauer, M. (2020). Leveraging multilingual identities in computer science education. In M. R. Freiermuth & N. Zarrinabadi (Eds.), *Of second language learners and users*. Palgrave-Macmillan. https://doi.org/10.1007/978-3-030-34212-8_12

Jacob, S., & Warschauer, M. (2018). Computational thinking and literacy. *Journal of Computer Science Integration, 1*, 1–9.

Kamberelis, G., Gillis, V. R., & Leonard, J. (2014). Disciplinary literacy, English learners, and STEM education. *Action in Teacher Education, 36*(3), 187–191.

LaCosse, J., Canning, E. A., Bowman, N. A., Murphy, M. C., & Logel, C. (2020). A social-belonging intervention improves STEM outcomes for students who speak English as a second language. *Science Advances, 6*(4), eabb6543.

Ladson-Billings, G. (1995). But that's just good teaching! The case for culturally relevant pedagogy. *Theory into Practice, 34*, 159–165.

Lefever-Davis, S., & Pearman, C. J. (2015). Reading, writing and relevancy: Integrating 3R's into STEM. *The Open Communication Journal, 9*(1), 61–64.

Lowe, T. A., & Brophy, S. P. (2019). *Identifying computational thinking in storytelling literacy activities with scratch Jr.* 2019 ASEE Annual Conference & Exposition.

Luna, N. A., & Martinez, M. (2013). A qualitative study using community cultural wealth to understand the educational experiences of Latino college students. *Journal of Praxis in Multicultural Education, 7*(1), 2.

Maarouf, S. A. (2019). Supporting academic growth of English language learners: Integrating reading into STEM curriculum. *World Journal of Education, 9*(4), 83–96.

Owens, M., & Ramsay-Jordan, N. (2021). Diversity in STEM: A look at STEM choices amongst black and latinx high school students. *Journal of Underrepresented & Minority Progress, 5*(1), pp. 59-78.

Rivera, H. J., & Molina, R. (2017). Building literacy skills with early readers through stem activities. *Theme: Preparing Early Readers for Success, 52*(1), 20.

Shute, V. J., et al. (2017). Demystifying computational thinking. *Educational Research Review, 22*, 142–158.

Smith, A., et al. (2020). Toward a block-based programming approach to interactive storytelling for upper elementary students. In A.-G. Bosser, D.E. Millard, & C. Hargood (Eds.), *Interactive storytelling* (pp. 111–119). Springer.

Song, D., et al. (2021). Applying computational analysis of novice learners' computer programming patterns to reveal self-regulated learning, computational thinking, and learning performance. *Computers in Human Behavior, 120*, 106746.

Teng, F. (2022). Vocabulary learning through videos: Captions, advance-organizer strategy, and their combination. *Computer Assisted Language Learning, 35*(3), 518–550.

Tissenbaum, M., et al. (2019). From computational thinking to computational action. *Communications of the ACM, 62*(3), 34–36.

Popat, S., & Starkey, L. (2019). Learning to code or coding to learn? A systematic review. *Computers & Education, 128*, 365–376.

Wood, K., Jones, J., Stover, K., & Polly, D. (2011). STEM literacies: Integrating reading, writing, and technology in science and mathematics. *Middle School Journal, 43*(1), 55–62.

Chapter 4

Child-Robot Musical Theater for Diverse and Inclusive STREAMS Education for Young Children

Koeun Choi and Myounghoon Jeon

Science, Technology, Engineering, and Mathematics (STEM) education has gained significant attention due to its wide-ranging advantages, including preparing students for future career opportunities in STEM fields and promoting critical thinking and innovation (National Research Council, 2013). Subsequently, STEAM has emerged as an approach that complements STEM by integrating innovative and creative thinking processes, incorporating the arts (A) into STEM education (Robelen, 2011; Yakman, 2008). More recently, there has been an ongoing discourse on expanding this integration to include reading (STREAM; Maarouf, 2019) as well as the humanities and social science disciplines (Colucci-Gray et al., 2019). This comprehensive integration is evident in the scope of this book, which reflects STREAMS (Science, Technology, Reading, Engineering, Arts, Mathematics, and Social Sciences).

This chapter presents a potential approach to infusing STREAMS content into robotics education by leveraging the synergy of social robots and theater arts. Collaborative theater activities, co-created by children and social robots, provide a rich learning context in which children can explore the world, discover themselves, and develop an understanding of others through their interactions with these robots. This chapter introduces a child-robot theater program designed to incorporate the diverse interests and experiences of children. The program systematically integrates different artistic formats, interactive storytelling, and collaborative performances involving both children and social robots. Specifically, it illustrates a child-robot musical theater program tailored to young children in an early childhood classroom setting. This chapter highlights the strategies used to systematically integrate the different facets of STREAMS while promoting inclusivity in the program, with the overarching goal of diversifying robotics education and promoting inclusivity.

LITERATURE REVIEW

Theoretical Frameworks

A large body of developmental and educational literature has underscored the active role young children play in exploring and learning about the world, others, and themselves throughout early childhood (Piaget, 1954; Gopnik, 2012). Piaget's constructivism emphasizes that children are naturally curious and driven to inquire and explore the world (Piaget, 1954). Many scholars have recognized similarities between young children's exploratory play and scientific endeavors in which children observe, hypothesize, and intervene to understand the world around them (Carey, 1985; Gopnik & Meltzoff, 1997; Wellman & Gelman, 1992). Thus, integrating children's genuine interests, natural curiosity, and everyday experiences into educational activities, including STEAM, is not only possible, but critical (Gopnik, 2012).

In the field of robotics education, Papert (1980) proposed constructionism, which builds on Piaget's constructivism but goes further by recognizing the role of technological tools in children's development. According to Papert (1980), the intentional design and use of technology provide a concrete way for children to explore abstract ideas once considered too complex for them to understand (Papert, 1980). Expanding on Papert's work, Bers (2012) introduced the Positive Technological Development (PTD) framework. This framework provides a comprehensive view of robotics education, taking into account children's developmental characteristics and holistic educational outcomes. According to the PTD framework, well-designed technological environments can promote children's interpersonal (competence, confidence, and character) and intrapersonal (caring, connection, and contribution) assets by fostering six positive behaviors, including collaboration, communication, community building, content creation, creativity, and choices of conduct (Bers, 2012).

Guided by the theoretical frameworks that emphasize young children's active learning (Piaget, 1954; Gopnik, 2012) and the role that technology plays in supporting the active learning process (Bers, 2012; Papert, 1980), a child-robot theater program introduced in this chapter was designed to integrate social robots into developmentally appropriate educational activities that enable children's active engagement in their learning process in order to promote children's positive behaviors and thereby facilitate their holistic development.

Social Robots and STREAMS

Social robots display physical and behavioral features to interact and communicate with humans, including facial expressions, gestures, body movements, and speech. These features allow social robots to create a sense of presence and social connection that increases engagement and facilitates natural interactions (Belpaeme et al., 2018). Social cues provided by technology have been proposed to support learning (Choi et al., 2022; Richert et al., 2011), with particular benefits for younger learners (Barr, 2019; Troseth, 2010). Indeed, social robots have been used to engage preschoolers and elementary school children in storytelling and educational games, leading to positive cognitive and affective outcomes (Papadopoulos et al., 2020).

Efforts have been made to integrate educational robotics kits into STREAMS, such as combining them with dance (Sullivan & Bers, 2018), acting and drawing (Sullivan et al., 2017), and storytelling (Yang et al., 2023) for young children. However, there are limitations to be addressed, including insufficient representations of robots (Darmawansah et al., 2023; Xia & Zhong, 2018) and the range of artistic formats (Sullivan & Bers, 2018). Moreover, prior attempts have predominantly focused on using robots as pedagogical tools for learning (Darmawansah et al., 2023), rather than recognizing their potential as social partners that support learners through social interaction.

Incorporating the social cues provided by robots may facilitate meaningful engagement for even younger children, thereby expanding the age range of children who can benefit from robotic education (Xia & Zhong, 2018). Furthermore, the integration of diverse social robots with a variety of artistic forms has the potential to encompass children's diverse interests and different avenues for artistic expression, thereby contributing to increasing diversity and inclusion. Nevertheless, integrating multiple subject areas and different robots in a cohesive and meaningful manner presents challenges. Given that not all interactions with technology have been shown to be beneficial for learning (Choi et al., 2021; Kirkorian, 2018), the intentional design and integration of interactive technologies is critical for educational benefits (Bers, 2012; Choi, 2021; Hirsh-Pasek et al., 2015). In this chapter, theater arts were considered a potential way to address the complexities of merging technology, creative arts, and education, with the goal of overcoming these challenges and fostering a more inclusive and enriching learning experience for young learners.

Social Robots and Theater Arts

Creating interactive theater programs with robots has been proposed as an approach to provide children with integrated and holistic learning experiences

(Bravo et al., 2017; Jeon et al., 2016). Theater arts encompass a range of artistic expressions, such as acting, dancing, music, and drawing, that are commonly observed in children's daily lives. Research has shown that educational activities paired with the arts promote the holistic development of young children, extending beyond arts-related topics (Rowe et al., 2018; Schnyder et al., 2021; Tominey & McClelland, 2011). Through child-robot theater activities, children can engage in multisensory learning to learn about the visual and auditory features, the physical appearance and movements, and the social-emotional experiences of humans and robots (Bravo et al., 2017). Through its storytelling component, theater activities also provide opportunities to integrate reading and social science topics. Thus, child-robot theater activities enable variable channels for acquiring information, constructing knowledge, and expressing creativity.

Theater activities can serve as a particularly rich context for the meaningful integration of a variety of artistic formats, as well as types of robots. Active engagement of young children in meaningful hands-on experiences that resonate with them has been shown to foster positive attitudes and beliefs about learning (Dejarnette, 2018; Master et al., 2017). Thus, theater activities involving collaborative performances co-created by children and social robots may serve as a developmentally appropriate approach to creating diverse and inclusive STEAM education by reflecting children's diverse interests and allowing for various creative expressions.

Indeed, there is a growing interest in integrating theater arts into human-robot interaction and social robots for children. Recent research suggests several educational benefits of such endeavors. Peng et al. (2020) found that children between the ages of 6–12 years enjoyed a live robot theater with multiple robotic agents and comprehended stories (Peng et al., 2020). Master et al. (2017) showed that even a brief 20-minute play activity of programming a dog robot significantly improved self-efficacy in 6-year-old girls. In a nine-week program for 4- to 6-year-old children with autism spectrum disorder, So et al. (2020) found that role-playing with robots enhanced children's engagement over the course of the program. Similarly, multi-week programs that include creative story development, prop and set design, rehearsals, and a final theatrical performance have been shown to play a positive role in kindergarten and elementary school children's interest and learning in STEAM subjects (Barnes et al., 2017, 2020; Dong et al., 2023).

The integration of diverse social robots and theater arts holds promise, yet bringing these emerging technological tools into educational contexts poses practical challenges. Appropriate educational programs and resources are pivotal in addressing these challenges. In response, this chapter presents a vignette featuring a child-robot musical theater program, tailored for young children in an early childhood classroom setting. This chapter outlines strategies for systematically integrating various aspects of STREAMS into early childhood classrooms for.

Educational Modules in Child-Robot Musical Theater Programs

Modular approaches have been frequently utilized in past child-robot musical theater programs (Barns et al., 2019, 2020; Jeon et al., 2016). These programs were organized into four modules focusing on four specific theater arts: acting, dancing, sound/music, and drawing. Each module was designed to help children learn about how to make robots act, dance, make sound, and draw, incorporating related STEM content. Simultaneously, these modules prepare the children for their final musical performance with social robots. Typically, each module was conducted either once or twice, either within a single day or spread out over two weeks. These modules provide flexibility, allowing the rearrangement of the module order and the option to skip a module. This format ensures continuity even if children miss a specific module. The program introduced as a vignette in this chapter also incorporated these four modules, each presented once per week.

Previous child-robot musical theater programs have been centered around unique storylines. While some programs have allowed participants to craft their own narratives, these approaches tend to be tailored for adolescents (Ko et al., 2020). When it comes to younger children, researchers have commonly provided a predetermined storyline while encouraging creativity. This is achieved by allowing children to assign and express characters through social robots or modify elements of the narratives (Dong et al., 2023). For example, existing stories, such as *Beauty and the Beast* or *Wizard of Oz*, are adapted for elementary school children. Instead of preassigning roles to social robots, researchers guide children to assign specific characters to robots based on the physical appearance and functionality of each robot. Alternatively, other programs created narratives featuring social robots, taking into account their specific functionalities (Choi et al., 2023). Crafting tailored stories requires additional effort but provides the benefit of enabling children to learn about the different characteristics of social robots within the context of the story. The latter approach is utilized in the program introduced in this chapter.

The narrative component in child-robot musical programs plays an important role in literacy education. Within these programs, children actively engage in and expand the story through questions, discussions, and extensions. They are also encouraged to creatively contribute to the story by transforming it into to a musical play script, which they practice and perform. Moreover, child-robot musical programs provide a platform for deliberately selecting stories that address specific social science topics, aligning closely with targeted educational objectives. For example, social robots have been integrated into an interactive musical theater performance centered around climate change as an educational outreach activity beyond the classroom (Lee et al., 2022). Equipped with advanced technologies, such as 140 speakers, a Cyclorama, and an optical motion tracking system, this immersive experience enabled the social robots not only to perform theatrical acts but also to engage the audience through contingent interactions. In these interactions, the social robots emphasized the significance of the audience's social responsibility in addressing environmental challenges. In the child-robot theater program introduced in the following vignette, the focus of the story was on various emotional experiences and coping strategies.

VIGNETTE: A CHILD-ROBOT MUSICAL THEATER PROGRAM FOR CHILDREN IN EARLY CHILDHOOD

Curriculum Overview

The vignette focuses on a child-robot musical theater program for young children in early childhood classrooms. The primary goal of this program was to provide a comprehensive classroom-based educational program for young children that systematically incorporates the social aspect of robots into early childhood education by integrating different types of social robots with theater arts, which encompass a variety of art formats. A seven-week child-robot theater program was designed for young children, with scheduled weekly visits. The program included four modules centered around different art themes (acting, dancing, sound/music, and drawing) and a musical theater performance. Additionally, both pre- and post-program visits were included for both initial preparations and final wrap-up activities.

Each module consisted of 30 minutes, comprising a 5-minute group storytime (see figure 4.1a), followed by 15-minute guided educational activities (see figure 4.1b), and concluding with 10-minutes of free play (see figure 4.1c). During the storytime segment, a researcher read the storybook, encouraging children to actively participate by answering questions about the narrative and engaging in singing and dancing along to the introduced song. The session then transitioned to educational activities, designed to be primarily hands-on, promoting children's exploration and enhancing their learning experiences. Finally, each module concluded with free play, where children had the opportunity to interact freely with multiple robots, under the supervision of both researchers and classroom teachers.

In the first week preceding the first module, researchers prepared the program by assessing the environment, ensuring the robots' internet connectivity

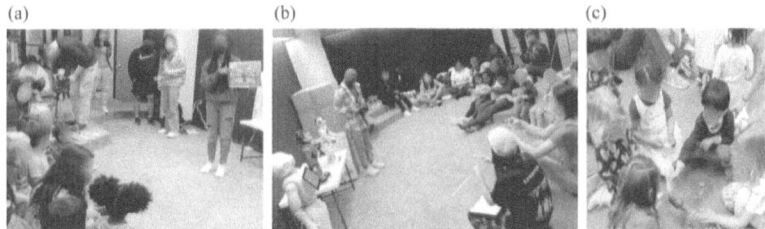

Figure 4.1 Example Module Activities Included (a) Storytime, (b) Educational Activities, and (c) Free Play. *Source*: K. Choi and M. Jeon (2022), Child Robot Theater Module Activities [Photograph], Virginia Tech Child Development Center for Learning and Research, Blacksburg, VA.

and power supply, and conducting interviews with children to assess their expectations and with teachers to gather their suggestions, concerns, and expectations. In the last week, after the final performance, researchers concluded the program by conducting follow-up interviews with children and teachers to gather their feedback on the program.

Educational Environment

The program was conducted in an empty room of a local childcare center in the southeastern region of the United States (seefigure 4.1). Researchers first set up the robots in the designated room. Each classroom participated in the program sequentially, visiting the room where the researchers were stationed. The program was offered to all three classrooms at the childcare center, running from March to May 2022. A total of 47 children (15–65 months) attended the program, with children from each classroom as a separate group. The classrooms had 12 children (17–36 months), 18 children (37–55 months), and 17 children (57–66 months), respectively. Before each module began, classroom teachers escorted children to the designated room. During each module, at least one classroom teacher was present to facilitate children's participation in educational activities. The room was equipped with a carpet where children could sit. Additionally, a smooth-surfaced table was provided to ensure clear visibility and mobility for the robots, except for Pepper, which was placed on the floor because of its size.

Social Robots in the Child-Robot Musical Theater Program

A diverse range of social robots have been incorporated into prior child-robot musical theater programs (Barns et al., 2017; Dong et al., 2023). Figure 4.2 depicts the social robots included in the current child-robot musical theater program for young children. Designed to resemble humans and animals,

Figure 4.2 The Social Robots Used in the Program. (a) Pepper, (b) NAO, (c) Milo, (d) Pleo, (e) Aibo, (f) Quincy. *Source*: K. Choi and M. Jeon (2022), Social Robots in the Child Robot Theater Program [Photograph], Blacksburg, VA.

these robots have the potential to enhance social interactions through speech, gestures, and body movements. The subsequent descriptions provide details about the physical characteristics and functions of these robots, as well as their applications within the child-robot musical theater program introduced in the chapter.

Pepper (see figure 4.2a; 48″ × 17″ × 19″) is a humanoid robot designed for social interactions. Developed by SoftBank Robotics, Pepper is equipped with advanced sensors and algorithms, enabling it to perceive and respond to human speech, gaze, and gestures. Pepper can engage in conversation through verbal and non-verbal means. It is capable of providing speech, adjusting its gaze, and making eye contact with the user during interaction. Additionally, its arms have a broad range of motions for expressive gestures. In the child-robot theater program, Pepper's speech and movements were controlled through Choregraphe, a desktop application offered by SoftBank.

NAO (see figure 4.2b; 22.6″ × 12.2″ × 10.8″) is a humanoid robot developed by SoftBank Robotics that resembles the human form with a head, torso, and limbs. It recognizes and responds to human speech and gestures. NAO is capable of speaking and performing a wide range of motions in its arms, legs, and head. In the child-robot theater program, NAO was programmed with advised speech and movements using Choregraph.

Milo (see figure 4.2c; 23.3″ × 7.5″ × 9.8″) is a humanoid robot developed by Robokind primarily for children with autism. Milo's facial and vocal expressions are programmable, allowing for a range of emotional expressions. Milo can move its arms and legs to walk and gesture. During the child-robot theater program, Milo's facial expressions and movements were programmed using Java. Further, PuTTY, an open-source terminal emulator, was used to control Milo through a serial port.

Pleo (see figure 4.2d; 7″ × 20″ × 12.7″) is an animal robot designed by Innvo Labs to resemble a dinosaur. Equipped with touch sensors covering its entire body, it can respond to user touches. Pleo produces dinosaur sounds and moves its body, limbs, head, and tail automatically. Additionally, Pleo can engage with specifically designed objects, such as toy leaves or stones, mimicking the act of eating or playing. During the child-robot theater program, Pleo operated in automatic mode without specific programming. That is, children interacted with Pleo freely, such as feeding it toy leaves and stroking its back.

Aibo (see figure 4.2e; 11.5″ × 12″ × 7.1″) is an animal robot developed by Sony. Aibo is a robot dog with a voice recognition feature, enabling it to recognize and respond to users' verbal commands. It produces dog-like sounds and conveys emotions through its eyes, ears, tail, and body movements. It is capable of autonomous movements and interactions with specifically designed objects, such as toy balls and bones, by moving or playing with them. Aibo can also recognize faces and adjust its position to approach accordingly. In the child-robot theater program, Aibo operated autonomously without specific programming.

Quincy (see figure 4.2f; 6″ × 4.1″ × 4.1″) is a drawing robot developed by Mindware. Quincy is equipped with arms where a pen can be attached and an eye that reads QR codes embedded in physical cards. These allow Quincy to offer step-by-step guidance for drawing various images through verbal instructions and precise movements of its arms. During the child-robot theater program, Quincy's existing features were used instead of programming additional gestures or speech.

Exploration of STREAMS Concepts in the Child-Robot Theater Program

Reading and Social Sciences in the Child-Robot Theater Program

For the purpose of this program, a storybook titled *Let's Go on a Walk*! was created using Canva, a graphical software tool (see figure 4.3). The storybook was designed to incorporate several robots featured in the program, namely Pepper, NAO, Milo, Pleo, and Aibo. The story centered around Milo, a robot who can express his emotions through facial expressions. In the story, Milo was sad, so he went for a walk with his dog, Aibo. Along the way, Milo encountered other robot friends, NAO and Pepper, who sang and danced, helping Milo navigate his emotional experiences. The storybook had questions to engage children (e.g., "Milo is feeling sad. What do you do when you feel sad?"). Each page of the digital storybook was printed at 11″ × 17″ dimensions and bound together to create a physical book. At the beginning of each module, a researcher read the storybook and encouraged children to

Figure 4.3 The Storybook Used in the Child-Robot Musical Theater Program for Young Children. *Source*: K. Choi and M. Jeon (2022), Example Storybook Pages in the Child Robot Theater Program [Photograph], Blacksburg, VA.

answer questions about the story and to participate in singing and dancing activities. Throughout these sessions, children actively engaged with the story, expanding it through questions, discussions, and extensions. Moreover, children were encouraged to contribute creatively to the story by transforming it into to a musical play script, which they then practiced and performed. This interactive approach fostered a dynamic and participatory learning environment for children.

Science, Technology, Engineering, Mathematics, and Arts in the Child-Robot Theater Program

The specific educational activities, focusing on the STEM aspects behind social robotics, were structured based on four theater art themes, all woven into the story. The educational activities were designed to be hands-on, beginning with a brief demonstration and instruction and then involving children in the implementation of the motions. Each module started with a large group storytime, led by one of the researchers. Following the reading, NAO, functioning as a robot instructor throughout the program, introduced the activities planned for the specific module. During the educational activities, one of the researchers operated NAO using the Choreograph application on her laptop, following the lesson scripts developed by the researchers. NAO's introduction was followed by researchers who demonstrated and explained details for the learning activities to the entire group. Subsequently, children were divided into small groups of 3–5 children, engaging in hands-on activities with robots under the guidance of a researcher or classroom teacher. The small group activities were combined with free play sessions for other groups, allowing them to wait for their turn to interact with the target robots while also enjoying playtime with the remaining robots.

Acting: In the acting module, children learned how to create facial and verbal expressions for robots through programming, controller buttons, and speech (seefigure 4.4). In a large group setting, a researcher demonstrated how to control Milo's facial and verbal expressions with a laptop using Java and controller buttons on the robot's body. This demonstration reflected children's suggestions and feedback on how to change Milo's facial features. In addition, the researcher explained how Aibo, the robot dog, can move and respond to children's speech input through sensors. Children engaged in discussions about the size and functionalities of these robots and explored how to create different emotional expressions through facial, vocal, and bodily movements using the robots. In small groups of 3–5 children, children were tasked with creating expressions for Milo and Aibo, enabling them to act out scenes from the storybook. Milo, as the main character, displayed various emotional states in the story, such as happiness and sadness, while Aibo, acting as Milo's pet, attempted to cheer up Milo. Some children worked on changing Milo's expressions by working with the researcher or pushing specific controller buttons. Other children worked on changing Aibo's bodily movements, employing voice commands.

Dancing: In the dancing module, children explored how to create robot movements and gestures using programming, speech, and sensors (see figure 4.5). The researcher demonstrated dance movements of NAO and Pepper, explaining concepts such as balance and how robots can mimic human actions. In the large group setting, children engaged in discussions around the similarities and differences between robot and human actions. Pleo, the dinosaur robot, was also introduced to show a range of responses to user input through sensors. In small groups, children were tasked with creating dance movements for Pepper, NAO, and Pleo, allowing them to dance based on the storybook narrative. In the storybook, NAO was portrayed as one of the characters capable of dancing, while Pleo was introduced as another friend that

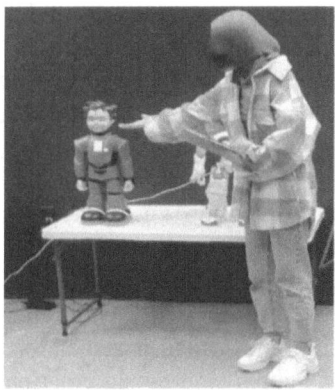

Figure 4.4 Examples of Educational Activities in the Acting Module. *Source*: K. Choi and M. Jeon (2022), Child Robot Theater Acting Module Activities [Photograph], Virginia Tech Child Development Center for Learning and Research, Blacksburg, VA.

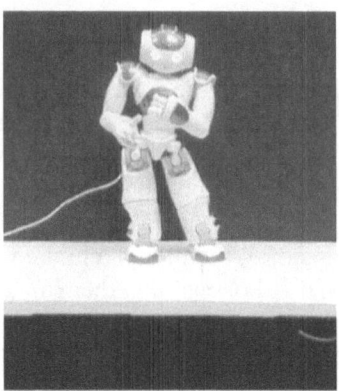 

Figure 4.5 **Examples of Educational Activities in the Dancing Module.** *Source*: K. Choi and M. Jeon (2022), Child Robot Theater Dancing Module Activities [Photograph], Virginia Tech Child Development Center for Learning and Research, Blacksburg, VA.

was encountered during his walk. Some children interacted with NAO, controlling its dance movements through a laptop operated by a researcher using the Choreograph application. Children asked questions and made requests to the researcher as they created specific dance movements for NAO. Other children danced with Pepper using its preprogrammed motions with the guidance of researchers. Children also had the opportunity to freely interact with Pleo by feeding and petting it, observing how Pleo moved its body in response.

Sound and Music: For the sound and music module, children learned about how to generate music and sound with technology, including artificial intelligence, synthesizers, and motion recognition (see figure 4.6). In the large group setting, a researcher provided demonstrations and explanations to introduce how technology could be used to generate sound effects and music. Pepper, the largest humanoid robot in the program, interacted with children to create music by offering an invisible piano keyboard by opening its arms. Children could move their hands to produce music through motion recognition. A music creation app, such as Groovepad–Music and Beat Maker, was introduced, enabling children to experiment with various musical elements and create music compositions. Further, an invention kit, Makey Makey, was used to create musical circuits that connected a computer and everyday objects such as fruits and vegetables. These objects functioned as keypads that could serve as bongo or piano keyboards, through which children generated electronic music. In small groups, children worked together to create background music and sound effects using Pepper, the music creation app, and musical circuits based on the events in the storybook. Within the storybook, Pepper was depicted as a character with the ability to sing. Some children interacted with Pepper, moving their hands to generate music, which

Pepper detected and transformed into corresponding sounds. Other children created music to convey Milo's emotions, expressing his moments of sadness and happiness using either the app or circuits.

Drawing: For the drawing module, children learned about the collaborative process of creating art with robots using digital image sensors (see figure 4.7). In a large group setting, a researcher introduced artworks generated by robots and showed how Quincy, a drawing robot, could scan QR codes embedded in physical cards using digital image sensors. Quincy then worked step by step to create corresponding drawings. Subsequently, eight Quincy robots

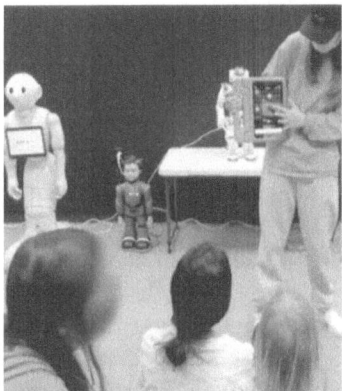

Figure 4.6 Examples of Educational Activities in the Sound and Music Module. *Source*: K. Choi and M. Jeon (2022), Child Robot Theater Sound and Music Module Activities [Photograph], Virginia Tech Child Development Center for Learning and Research, Blacksburg, VA.

Figure 4.7 Examples of Educational Activities in the Drawing Module. *Source*: K. Choi and M. Jeon (2022), Child Robot Theater Drawing Module Activities [Photograph], Virginia Tech Child Development Center for Learning and Research, Blacksburg, VA.

Figure 4.8 Example of the Final Performance. *Source*: K. Choi and M. Jeon (2022), Child Robot Theater Final Performance [Photograph], Virginia Tech Child Development Center for Learning and Research, Blacksburg, VA.

were distributed among the children. In smaller groups, children interacted with Quincy to draw images inspired by the storybook, where Milo and his robot friends enjoyed a walk and play in the park. Children selected images by choosing cards with specific illustrations (e.g., cat, fox, or flower) and embedded QR codes. Some children extended Quincy's drawing by coloring or adding additional elements. Meanwhile, other children created their own drawings for the background scenery. The collective artworks served as the backdrop for the final performance.

Performance: After completing the four modules, children from each classroom presented a child-robot musical theater performance incorporating all the robots and elements from the four modules (figure 4.8). The musical theater performance was interactive, engaging children by encouraging their verbal responses and active participation through joint singing and dancing along with the story. The performance was attended by children's parents and classroom teachers, creating dynamic and participatory experiences for everyone involved.

DISCUSSION

STREAMS through Social Robots and Theater Arts

The vignette highlights the promising potential of incorporating social robots into theater activities to enhance early childhood STREAMS education. The collaborative process of creating the theater play by children and social robots empowers children to actively construct their own learning experiences (Bers, 2012; Chi, 2009; Papert, 1980; Piaget, 1954). Through a meaningful and coherent integration of diverse domains and artistic elements, child-robot theater programs provide a rich context that reflects children's diverse interests and allows for a range of expressions. These educational activities offer the opportunity to effectively weave together children's wide-ranging interests, natural curiosity, and everyday experiences into their learning process (Gopnik, 2012).

In the child-robot musical theater program, social robots serve as social partners engaging children in educational activities. These interactions offer valuable opportunities for children to enhance their developmental processes through observation, imitation, direct instruction, and contingent interaction (Lillard et al., 2013; Vygotsky, 1978). Social cues presented by technologies can evoke social responses from children, which may also shape their perception of these tools as social entities (Melson et al., 2009; Richert et al., 2011). The deliberate integration of social cues provided by robots is expected to enhance meaningful engagement, especially yielding greater benefits for younger learners (Barr, 2019; Troseth, 2010). Thus, harnessing the social aspects of robotic technology holds significant potential to broaden the age range of children who can benefit from education involving robots.

In the child-robot musical theater program, social robots portray various roles as actors, expanding their conventional positions as instructors, peers, or novices (Belpaeme et al., 2018). Their distinct roles provide children with unique opportunities to engage with technology by exploring imaginative scenarios as well as collaborating and emphasizing with different characters depicted by robots. The ontology of social robots may inspire comparisons between social robots, mechanical robots, and human partners among children (Weisman, 2022). Such analyses lead to inquiries into various dimensions of human-computer interaction, encompassing biological, cognitive, social, emotional, and moral domains. Thus, the diverse roles fulfilled by social robots in child-robot musical theater programs hold the promise of enhancing children's learning experiences through improving social and cognitive skills and stimulating creativity.

Learning Environments and Educator Roles

The child-robot musical theater program has been successfully implemented in early childhood classrooms. The program was designed to provide flexibility, allowing for adaptability in terms of overall duration, module length and order, and location. The program also enabled the customization of age-appropriate content and instructional tools, which include physical books and visual aids. Social robots and related technologies were introduced and brought to the school by researchers, necessitating minimal physical resources from the participating institutions, such as internet connectivity and power supply for the robots.

The program entailed collaboration between multiple researchers, teachers, and staff. Active engagement from classroom teachers was crucial in sustaining children's engagement while providing guidance during large and small group activities and free play sessions. In this program, all of the children from each classroom were included to promote equity. In this context, teachers played a vital role in addressing the diverse needs of children by offering sufficient attention, regulating behaviors, and fostering engagement at an individual level.

In the child-robot theater program, researchers took the lead role in the educational activities, being responsible for bringing and managing all robots throughout the programs. This approach eased the burden on the teachers, giving them the opportunity to interact with children and support their learning. We believe that our programs served as crucial building blocks for everyone involved, including teachers. As teachers actively engaged in the program and supported children to interact with the social robots, they had the opportunity to acquire essential skills and gain insights to develop educational activities incorporating social robots into their teaching. The important next step is to understand how to support teachers as the primary users of the social robots, including the development of assistive tools and educational resources.

FUTURE DIRECTIONS

Many of the social robots used in the child-robot musical theater program were high-priced and specialized for research purposes. This necessitated the use of various software tools and considerable training. Given the diverse range of social robots available, it is crucial to explore more affordable models to enhance accessibility. Additionally, developing training materials and educational resources for teachers is critical to scaling up these initiatives.

Given the frequent use of robotic kits within educational robotics, it would be valuable to consider whether and how social robots and theater arts could be integrated with these kits. Social aspects have been incorporated into robotic kits through children's creative contributions, such as assembling blocks or adding crafted materials, as demonstrated in studies involving LEGO (Bers et al., 2014), KIWI (Sullivan & Bers, 2016), and KIBO (Sullivan & Bers, 2018). Investigating innovative ways to integrate these robotic toolkits with social robots and theater arts activities would not only utilize the problem-solving and creativity aspects of robotics kits but also tap into the interactive and expressive potential of social robots and theater arts. The synergy of these components could result in a more enriching educational experience, promoting holistic learning for young learners.

The vignette lays the groundwork for refining STREAMS curricula that incorporate social robots, serving as an important starting point. To enhance this foundation, it is crucial to develop specific and tailored learning objectives and activities for each domain. For example, the vignette suggests that a child-robot theater program could provide a meaningful context for the in-depth exploration of social science topics. Analyzing characters and storylines within theater activities requires children to employ critical thinking skills. By immersing students in the emotions, struggles, and dilemmas of diverse characters, theater activities can be used to cultivate empathy and encourage contemplation of ethical issues from multiple viewpoints. These discussions can be further enriched with cultural and historical perspectives. Child-robot theater activities provide a safe space for discussions on themes such as technology-related ethics, inclusion, and diversity. In future research, it is critical to systematically integrate multiple domains in learning objectives and educational activities and to assess the effectiveness of the child-robot musical theater program in learning outcomes across a wide range of subjects. This line of work will contribute to the continuous improvement of pedagogical approaches for interactive theater experiences with socially contingent technologies, ensuring a well-rounded and inclusive learning environment for children.

CONCLUSION

Integrating social robots into theater activities that involve child-robot social interactions can offer a valuable context for early childhood and elementary STREAM education. By exploring diverse social robots and incorporating various artistic forms, the vignette underscores the significance of embracing diversity and inclusivity. Further, the vignette highlights the need for further investigation into the ways social robots and theater arts can be effectively leveraged to support young children's holistic development. In light of the increasing integration of technology and human interaction, this line of work will contribute to broadening our understanding of the role of social robots in early childhood education.

REFERENCES

Barnes, J., FakhrHosseini, M. S., Vasey, E., Duford, Z., & Jeon, M. (2017). Robot theater with children for STEAM education. *Proceedings of the Human Factors and Ergonomics Society Annual Meeting*, *61*(1), 875–879. https://journals.sagepub.com/doi/pdf/10.1177/1541931213601511

Barnes, J., FakhrHosseini, S. M., Vasey, E., Park, C. H., & Jeon, M. (2019, May). Informal STEAM education case study: Child-robot musical theater. In *Extended Abstracts of the 2019 CHI Conference on Human Factors in Computing Systems* (pp. 1–6). ACM Press, Glasgow, UK.

Barnes, J., FakhrHosseini, S. M., Vasey, E., Park, C. H., & Jeon, M. (2020). Child-robot theater: Engaging elementary students in informal STEAM education using robots. *IEEE Pervasive Computing, 19*(1), 22–31. https://doi.org/10.1109/MPRV.2019.2940181

Barr, R. (2019). Growing up in the digital age: Early learning and family media ecology. *Current Directions in Psychological Science, 28*(4), 341–346. https://doi.org/10.1177/0963721419838245

Belpaeme, T., Kennedy, J., Ramachandran, A., Scassellati, B., & Tanaka, F. (2018). Social robots for education: A review. *Science Robotics, 3*(21), 5954. https://doi.org/10.1126/scirobotics.aat5954

Bers, M. U. (2012). *Designing digital experiences for positive youth development: From playpen to playground.* Oxford University Press.

Bers, M. U., Flannery, L., Kazakoff, E. R., & Sullivan, A. (2014). Computational thinking and tinkering: Exploration of an early childhood robotics curriculum. *Computers & Education, 72,* 145–157. http://dx.doi.org/10.1016/j.compedu.2013.10.020

Braun, V., & Clarke, V. (2012). *Thematic analysis.* American Psychological Association.

Bravo, F. A., González, A. M., & González, E. (2017, October). A review of intuitive robot programming environments for educational purposes. In *2017 IEEE 3rd Colombian Conference on Automatic Control (CCAC)* (pp. 1–6). IEEE. https://doi.org/10.1109/CCAC.2017.8276396

Carey, S. (1985). Are children fundamentally different kinds of thinkers and learners than adults. *Thinking and Learning Skills, 2,* 485–517. https://doi.org/10.4324/9780203056646-31

Chi, M. T. (2009). Active-constructive-interactive: A conceptual framework for differentiating learning activities. *Topics in Cognitive Science, 1*(1), 73–105. -8765.2008.01005.xCitations:%20767

Choi, K. (2021). Sesame street: Beyond 50. *Journal of Children and Media, 15*(4), 597–603. https://doi.org/10.1080/17482798.2021.1978675

Choi, K., Kirkorian, H. L., & Pempek, T. A. (2021). Touchscreens for whom? Working memory and age moderate the impact of contingency on toddlers' transfer from video. *Frontiers in Psychology, 12,* 621372. https://doi.org/10.3389/fpsyg.2021.621372

Choi, K., Schlesinger, M. A., Franchak, J. M., & Richert, R. A. (2022). Preschoolers' attention to and learning from on-screen characters that vary by effort and efficiency: An eye-tracking study. *Frontiers in Psychology, 13.* https://doi.org/10.3389/fpsyg.2022.1011172

Choi, K., Yu, S., Kim, J., Dong, J., Lee, Y., Haines, C., Newbill, P., Upthegrove, T., Wyatt, A., & Jeon, M. (2023, April 13–16). *Interactive stories through robot musical theater for preschoolers' STEAM education* [Paper presentation]. The annual meeting of the American Educational Research Association, Chicago, IL.

Colucci-Gray, L., Burnard, P., Gray, D., & Cooke, C. (2019). *A critical review of STEAM (science, technology, engineering, arts, and mathematics)*. Oxford Research Encyclopedia of Education.

Darmawansah, D., Hwang, G. J., Chen, M. R. A., & Liang, J. C. (2023). Trends and research foci of robotics-based STEM education: A systematic review from diverse angles based on the technology-based learning model. *International Journal of STEM Education, 10*(1), 1–24. https://doi.org/10.1186/s40594-023-00400-3

Dejarnette, N. K. (2018). Implementing STEAM in the early childhood classroom. *European Journal of STEM Education, 3*(3), 1–9. https://doi.org/10.20897/ejsteme/3878

Dong, J., Choi, K., Yu, S., Lee, Y., Kim, J., Vajir, D., & Jeon, M. (2023): A child-robot musical theater afterschool program for promoting STEAM education: A case study and guidelines. *International Journal of Human–Computer Interaction, 40*(13), 3465–3481. https://doi.org/10.1080/10447318.2023.2189814

Gopnik, A. (2012). Scientific thinking in young children: Theoretical advances, empirical research, and policy implications. *Science, 337*(6102), 1623–1627. https://doi.org/10.1126/science.1223416

Gopnik, A., & Meltzoff, A. N. (1997). *Words, thoughts, and theories*. MIT Press.

Jeon, M., FakhrHosseini, M., Barnes, J., Duford, Z., Zhang, R., Ryan, J., & Vasey, E. (2016, March). Making live theater with multiple robots as actors bringing robots to rural schools to promote STEAM education for underserved students. In *2016 11th ACM/IEEE International Conference on Human-Robot Interaction (HRI)* (pp. 445–446). IEEE.

Kirkorian, H. L. (2018). When and how do interactive digital media help children connect what they see on and off the screen? *Child Development Perspectives, 12*(3), 210–214. https://doi.org/10.1111/cdep.12290

Ko, S., Swaim, H., Sanghavi, H., Dong, J., Nadri, C., & Jeon, M. (2020). Robot-theater programs for different age groups to promote STEAM education and robotics research. *Companion of the 2020 ACM/IEEE International Conference on Human-Robot Interaction* (pp. 299–301). https://doi.org/10.1145/3371382.3378353

Lee, Y., Wyatt, A., Dong, J., Upthegrove, T., Hale, B., Lyles, C. H., & Jeon, M. (2022, March). *Robot musical theater for climate change education*. In Proceedings of the 2022 ACM/IEEE International Conference on Human-Robot Interaction (pp. 870–874). https://trim.mtu.edu/publications/HRI2022.pdf

Lillard, A. S., Lerner, M. D., Hopkins, E. J., Dore, R. A., Smith, E. D., & Palmquist, C. M. (2013). The impact of pretend play on children's development: A review of the evidence. *Psychological Bulletin, 139*(1), 1. https://doi.org/10.1037/a0029321

Maarouf, S. A. (2019). Supporting academic growth of English language learners: Integrating reading into STEM curriculum. *World Journal of Education, 9*(4), 83–96. https://doi.org/10.5430/wje.v9n4p83

Master, A., Cheryan, S., Moscatelli, A., & Meltzoff, A. N. (2017). Programming experience promotes higher STEM motivation among first-grade girls. *Journal of Experimental Child Psychology, 160*, 92–106. https://doi.org/10.1016/j.jecp.2017.03.013

Melson, G. F., Kahn, P. H., Jr., Beck, A., & Friedman, B. (2009). Robotic pets in human lives: Implications for the human-animal bond and for human relationships with personified technologies. *Journal of Social Issues, 65*(3), 545–567. https://doi.org/10.1111/j.1540-4560.2009.01613.x

National Research Council. (2013). *Next generation science standards: For States, by States.* The National Academies Press. https://doi.org/10.17226/18290

Papadopoulos, I., Lazzarino, R., Miah, S., Weaver, T., Thomas, B., & Koulouglioti, C. (2020). A systematic review of the literature regarding socially assistive robots in pre-tertiary education. *Computers & Education, 155,* 103924. https://doi.org/10.1016/j.compedu.2020.103924

Papert, S. (1980). *Mindstorms: Children, computers, and powerful ideas.* Basic Books, Inc. http://dl.acm.org/citation.cfm?id1/41095592

Peng, Y., Feng, Y. L., Wang, N., & Mi, H. (2020, August). How children interpret robots' contextual behaviors in live theatre: Gaining insights for multi-robot theatre design. In *2020 29th IEEE International Conference on Robot and Human Interactive Communication* (RO-MAN) (pp. 327–334). IEEE.

Piaget, J. (1954). *The construction of reality in the child.* Basic Books. https://doi.org/10.1037/11168-000

Richert, R. A., Robb, M. B., & Smith, E. I. (2011). Media as social partners: The social nature of young children's learning from screen media. *Child Development, 82*(1), 82–95. https://doi.org/10.1111/j.1467-8624.2010.01542.x

Robelen, E. W. (2011). STEAM: Experts make case for adding arts to STEM. *Education Week, 31*(13), 8. https://www.edweek.org/ew/articles/2011/12/01/13steam_%20ep.h31.html

Rowe, M. L., Salo, V. C., & Rubin, K. (2018). Toward creativity: Do theatrical experiences improve pretend play and cooperation among preschoolers? *American Journal of Play, 10*(2), 193–207. https://eric.ed.gov/?id=EJ1179965

Schnyder, S. S., Wico, D. M., & Huber, T. (2021). Theater arts as a beneficial and educational venue in identifying and providing therapeutic coping skills for early childhood adversities: a systematic review of the literature. *International Electronic Journal of Elementary Education, 13*(4), 457–467. https://doi.org/10.26822/iejee.2021.204

So, W. C., Cheng, C. H., Lam, W. Y., Huang, Y., Ng, K. C., Tung, H. C., & Wong, W. (2020). A robot-based play-drama intervention may improve the joint attention and functional play behaviors of Chinese-speaking preschoolers with autism spectrum disorder: A pilot study. *Journal of Autism and Developmental Disorders, 50*(2), 467–481. https://doi.org/10.1007/s10803-019-04270-z

Sullivan, A., & Bers, M. U. (2016). Robotics in the early childhood classroom: Learning outcomes from an 8-week robotics curriculum in pre-kindergarten through second grade. *International Journal of Technology and Design Education, 26*(1), 3–20. https://doi.org/10.1007/s10798-015-9304-5

Sullivan, A., & Bers, M. U. (2018). Dancing robots: Integrating art, music, and robotics in Singapore's early childhood centers. *International Journal of Technology and Design Education, 28,* 325–346. https://doi.org/10.1007/s10798-017-9397-0

Sullivan, A., Strawhacker, A., & Bers, M. U. (2017). Dancing, drawing, and dramatic robots: Integrating robotics and the arts to teach foundational STEAM concepts to young children. In M. S. Khine (Ed.), *Robotics in STEM education: Redesigning the learning experience* (pp. 231–260). Springer.

Sung, J., Lee, J. Y., & Chun, H. Y. (2023). Short-term effects of a classroom-based STEAM program using robotic kits on children in South Korea. *International Journal of STEM Education, 10*(1), 1–18. https://doi.org/10.1186/s40594-023-00417-8

Troseth, G. L. (2010). Is it life or is it Memorex? Video as a representation of reality. *Developmental Review, 30*(2), 155–175. https://doi.org/10.1016/j.dr.2010.03.007

Tominey, S. L., & McClelland, M. M. (2011). Red light, purple light: Findings from a randomized trial using circle time games to improve behavioral self-regulation in preschool. *Early Education & Development, 22*(3), 489–519. https://doi.org/10.1080/10409289.2011.574258

Vygotsky, L. S. (1978). *Mind in society: The development of higher psychological processes.* Harvard University Press.

Weisman, K. (2022). Extraordinary entities: Insights into folk ontology from studies of lay people's beliefs about robots. In *Proceedings of the Annual Meeting of the Cognitive Science Society* (Vol. 44, No. 44). Toronto, Canada.

Wellman, H. M., & Gelman, S. A. (1992). Cognitive development: Foundational theories of core domains. *Annual Review of Psychology, 43*(1), 337–375. https://doi.org/10.1146/annurev.ps.43.020192.002005

Xia, L., & Zhong, B. (2018). A systematic review on teaching and learning robotics content knowledge in K–12. *Computers & Education, 127*, 267–282. https://doi.org/10.1016/j.compedu.2018.09.007

Yakman, G. (2008). STEAM education: An overview of creating a model of integrative education. In *Pupils' Attitudes Towards Technology (PATT-19) Conference: Research on Technology, Innovation, Design & Engineering Teaching*, Salt Lake City, Utah.

Yang, W., Ng, D. T. K., & Su, J. (2023). The impact of story-inspired programming on preschool children's computational thinking: A multi-group experiment. *Thinking Skills and Creativity, 47*, 101218. https://doi.org/10.1016/j.tsc.2022.101218

Chapter 5

Critical Service-Learning
A Transformational Framework for Integrating STEM and Social Justice
Cara Eleonora Daza

This chapter delves into the intricate disparities faced by women and underrepresented students, such as Black, Latinx, Indigenous, and first-generation students, in STEM. While a significant portion of the research primarily focuses on the United States, it is vital to recognize that these challenges transcend national borders, resonating on a global scale. Beyond these groups, other marginalized communities in STEM grapple with systemic inequalities rooted in factors like race, class, sexual orientation, and ability. The compounding effect of intersectionality intensifies these challenges. Continued research, dedicated spaces for listening and reflection, and proactive advocacy are essential for understanding and addressing the underrepresentation of all marginalized groups.

STEM and STREAMS education possesses the remarkable potential to address the world's most pressing issues while simultaneously addressing oppressive systems. By integrating social and environmental justice principles into curricula, STEM educators can not only attract and retain more women and underrepresented students but also cultivate diverse STEM graduates who are deeply committed to advancing a more equitable world.

This integration hinges on making the connections between content and context, inquiry and action, knowledge and responsibility *clear and relevant to learners*. To achieve this, an integrative framework known as STREAMS (Science, Technology, Reading/ELA, Engineering, Art, Mathematics, and Social Sciences) emerges as the vanguard of STEM education's evolution. STREAMS transcends the conventional boundaries of STEM by integrating a wider spectrum of disciplines, including social justice.

Within this framework, this chapter explores critical service-learning (CSL), a pedagogical approach that prioritizes relational, equitable, and action-oriented principles, integrating STEM and social justice to foster transformative educational experiences. Beginning with an exploration of the pervasive challenges faced by women and underrepresented students in STEM education, the chapter underscores the serious need for innovative solutions to address equity disparities. It then examines the efficacy of critical service-learning as a tool for promoting equity within STEM education, drawing on research to illustrate its impact. Furthermore, the chapter shows a real-world example of CSL implementation, alongside an exploration of the distinctions between traditional service-learning and critical service-learning. This examination aims to foster authentic equity work and mitigate the risk of perpetuating oppressive systems.

Inspiration for navigating this terrain comes from the works of esteemed scholars like Tania D. Mitchell, Paulo Freire, and bell hooks. Moreover, the concepts and insights regarding CSL presented in this chapter have been collaboratively developed in partnership with Colombian students, teachers, and community members, with whom I've had the honor of closely working.

POSITIONALITY

As a white, cisgender female from the American Midwest, my journey has led me to reside in Colombia, South America, for over a decade. My core values revolve around faith, family, and justice. When I initially embarked on my service-learning path, I inadvertently contributed to harm through white supremacy culture, such as thinking "the way I do things in the U.S. is the right way to do things," or power hoarding—making decisions *for* people, not *with* people. However, as I developed critical consciousness, I recognized my role in not only serving but also challenging and dismantling oppressive systems with love and empathy. Moreover, I believe it is unjust to place the burden of fostering equitable and just communities solely on underrepresented and underserved individuals. My commitment lies in inspiring educators, delivering quality education for all learners, particularly those who are marginalized, and advocating for the integration of critical service-learning in STEM education to drive transformative change in our communities.

Meet Millard

Millard McElwee's journey into the world of engineering was marked by purpose and resilience. At the age of 12, Hurricane Katrina struck his Louisiana hometown. Though he evacuated before the disaster, the devastation he witnessed when he returned made a profound impression. While his own home escaped extreme flooding, the residences of his relatives in New Orleans's Ninth Ward and Metairie were severely affected. Several relatives sought refuge in his two-bedroom home and endured one month without electricity, an experience that was etched into his memory as something he would never forget.

Another moment that left a lasting impact on Millard was the arrival of engineering experts from the National Institute of Standards and Technology and the University of California, Berkeley, in the aftermath of the natural disaster. They had come to Louisiana to assess what had gone wrong, and these visits sparked meaningful conversations between Millard and his father, who was also an engineer. Together, they talked about the reasons behind the failures of both the levees and the Army Corps of Engineers. These discussions ignited a newfound interest in Millard, propelling him toward a growing determination to become an engineer and contribute to the protection of vulnerable communities against future catastrophes.

As he navigated his educational journey, Millard's interest in engineering continued to grow. Despite not having access to high-quality STEM classes in his high school or mentor teachers who encouraged him in his pursuit of engineering, he dedicated himself to excelling in school and actively sought out learning opportunities within his community.

Volunteering at a local African American history museum also played a significant role in his academic development and identity. In high school, he spent countless hours exploring online resources and engineering opportunities, eventually discovering MIT's Minority Introduction to Engineering and Science program. Millard, a Black student, attended this program, and it solidified his desire to pursue an engineering degree.

Following high school, Millard was awarded the prestigious Gates Millennium Scholarship and pursued a degree in civil engineering at Carnegie Mellon University. Subsequently, he went on to attain a PhD from the University of California, Berkeley, with a specific research focus on quantifying the impact of natural disasters on communities of color.

His journey in higher education was marked by substantial accomplishments, but it was far from easy. Millard often found himself feeling isolated and lacking the necessary support in his engineering research projects related to social and environmental justice. During this period, he faced the challenge of being one of the few engineering students of color, with a notable absence of engineering professors from similar backgrounds. Furthermore, his exposure to ideas and scholarly materials was predominantly limited to those authored by white males.

Today, Millard McElwee is an engineering consultant specializing in projects related to environmental justice. He also serves as an adjunct professor at Santa Clara University School of Engineering and a project instructor at Carnegie Mellon University and Rice University. At the core of his story lies a deep dedication to giving back, as he aptly articulates it: "That sense of duty to pay it forward, knowing that I am standing on the shoulders of those who came before me, is what drives my work." Millard's motivation is rooted in addressing the underrepresentation of African Americans in STEM education and advancing the cause of environmental justice. It exemplifies his aspiration to contribute meaningfully to society, propelling his relentless endeavors to promote diversity in STEM education (McElwee, 2023).

STEM EDUCATION: ASSETS AND CHALLENGES

Over the past few decades, STEM education has evolved beyond a curriculum into a dynamic and meaningful learning experience, engaging both students and educators alike. Students are validated to unleash their curiosity, creativity, collaborate effectively, and employ critical thinking skills. Crucially, STEM education places a strong emphasis on real-world learning, interdisciplinary projects, and prepares students for the careers of the future, making it a global educational practice. The significance of STEM education is exemplified by the remarkable 79 percent increase in STEM occupation employment in the United States over the last three decades (O'Rourke, 2021). However, the value of STEM education extends far beyond employment. It equips future leaders to address the world's most pressing challenges.

While Millard's STEM journey serves as both inspiration and motivation, it's crucial to recognize that his experience as an individual of color does not reflect the typical STEM experience. The persistent issue of underrepresentation in STEM education has persisted for decades, particularly affecting women and marginalized minority groups, including Black, Latinx, Indigenous, and first-generation students.

The persistent disparities in STEM education, often referred to as "opportunity gaps," are long-standing issues that reflect deeply rooted systemic racism and gender disparities, both in the United States and internationally. These challenges, with their historical origins, pose a significant obstacle as they continue to constrain the range of perspectives, life experiences, and innovative ideas within the STEM domains, impeding equity and innovation. To quote Kenneth Gibbs, "When we view scientific research as a form of group problem-solving rather than the showcase of individual brilliance, diversity becomes indispensable for achieving excellence (Gibbs, 2014)." The crucial inquiry emerges: Despite decades of recognition, research, and discourse, why do these STEM disparities endure, and what strategies can effectively address them?

The reality is that these issues stem from a multifaceted interplay of factors and complex systems. These interconnected challenges persistently hinder progress toward equity and social justice, ultimately preventing students from realizing their full potential.

What is happening in the K–12 U.S. Educational system?

Lack of Engagement

Both females and minority groups often find science and mathematics unattractive in K–12 settings, perceiving the topics as irrelevant to their lives and disconnected from their existing knowledge (Finkel, 2018). Gender disparities also persist, with women receiving less exposure and guidance in STEM classes and concerning careers, compared to their male counterparts, impacting their aspirations to "intend" to major in STEM fields before higher education (Costello, Salehi, Ballen, & Burkholder, 2023). Stereotypes related to race and gender continue to permeate the education system, exerting a significant influence on students' experiences. For instance, a study on mathematics biases revealed the presence of both conscious and unconscious biases among teachers. Some teachers wrongly assume that boys possess greater mathematical ability than girls and that white students outperform students of color in mathematics (Copur-Gencturk, Thacker, & Cimpian, 2023). Additionally, educational research predominantly focuses on the ways in which students of color differ from white students within the context of schooling. This deficit-oriented perspective often inadvertently perpetuates stereotypes, reinforcing oppression rather than fostering meaningful change.

Furthermore, the limited diversity among classroom teachers is a prevailing issue, with just 28 percent from racially diverse backgrounds (Hugh-Pennie, Hernandez, Uwayo, Johnson & Ross, 2021). This lack of diversity poses significant challenges for minority students. Research demonstrates that students of color tend to receive lower behavioral and academic ratings when taught by white teachers, and they are also more likely to face stricter and heightened disciplinary measures, including referrals, suspensions, and expulsions, compared to their white peers (Hugh-Pennie et al., 2021). Many educational pedagogies often ignore or reinforce a culture that has actively excluded, oppressed, and even abused women and minorities. An example of this is the chronic underrepresentation of marginalized minority groups and women role models in the STEM curriculum, which hinders students' ability to envision themselves in STEM careers. Another example lies in the issue of traditional science learning experiences failing to respect Indigenous worldviews, putting students in a position where they must compromise their cultural identity in order to fit into the mainstream scientific community (Jin, 2012).

PURSUING SOLUTIONS

Diversity, Equity, and Inclusion

Diversity, Equity, and Inclusion (DEI) efforts are increasingly prominent in STEM education, with a growing number of educators, departments, and organizations prioritizing equity and social justice in their values and curricula. This shift is fueled by a pressing need to address inequities and bolstered by a mounting body of research at the intersection of STEM and social justice, driving the integration of these principles into K–12 and higher education STEM curricula. DEI training and equity-centered pedagogies like culturally relevant pedagogy are recognized for their role in enhancing student success (Hugh-Pennie et al., 2021). Organizations such as Learning for Justice provide valuable support to K–12 schools through resources, research initiatives, and professional development opportunities, fostering inclusive and equitable educational environments.

While progress in DEI within STEM education is vital, significant challenges persist. Achieving meaningful equity work requires full stakeholder engagement and genuine commitment, which can be complex and often leads some institutions to adopt a superficial approach. Moreover, navigating equity work within the educational landscape is demanding, given the daily rigors placed upon educators and the looming specter of burnout resulting from ever-increasing pressures, initiatives, and workloads. Without adequate training, support, opportunities for reflection, and access to an equity-driven curriculum, it is unrealistic to expect educators to embrace and champion principles of decolonization, anti-racism, anti-sexism, and anti-oppression.

Self-Reflection

To advance equity in STEM education, the often-overlooked yet essential solution is self-reflection within the STEM community. Educators must be willing to self-reflect on their practices and the systems they work within. Vincent Basile and Flávio Azevedo exemplify self-reflection in their article, "Ideology in the Mirror: A Compassionate Self-Examination of Our Efforts for Equity and Social Justice in STEM Education (Basile & Azevedo, 2022)." Their introspection into their STEM practices illuminates the challenges they've encountered and continue to face, hindering their pursuit of equity in STEM, including issues like oppression, capitalism, profiteering, racism, and militarism. They underscore how capitalism influences research, teaching methods, and the prioritization of certain subjects, perpetuating complacency in the pursuit of more equitable education models. They call upon STEM educators with a heartfelt challenge to contemplate the significance of committing to an ongoing process of ideological development and growth in equity work.

Furthermore, Basile and Azevedo (2021) shed light on the complexity of equity work, emphasizing that the dynamics within classrooms are profoundly shaped by teachers' beliefs, ideologies, and thought processes. "Conceptual change" denotes the acquisition of challenging knowledge that, with time and exposure, becomes comprehensible, as in science or math. However, when it comes to personal ideologies and beliefs, heavily influenced by our families, culture, and environment, individuals tend to make sense of the world in ways that can be intricate and inconsistent. This complexity is why both educators and students often grapple with discussing topics such as race, ableism, gender roles, and oppression. These subjects can trigger internal conflicts because conflicting ideas coexist within individuals, often without their awareness.

However, equity work demands educators to courageously confront cognitive dissonance. Encouraging self-reflection, questioning current beliefs, and embracing perspectives that transcend capitalism are essential steps. This entails scrutinizing funding sources, profit motives, and exploitation, while fostering acute awareness of oppression, actively listening to others, and compassionately vocalizing opposition against oppressive systems and behaviors.

Amidst the myriad complexities and disparities, it is tempting to assign blame, become apathetic, or even hopeless. However, rather than fixating on fault-finding, embracing an ethos of responsibility and hope emerges as a pathway to liberation. As bell hooks emphasizes, one of educators' responsibilities lies in actively engaging and wholeheartedly committing to the journey of self-actualization, underscored by prioritizing introspection and self-care (Specia & Osman, 2015). By cultivating these qualities within themselves and their students, educators lay the foundation for transformative classrooms. This steadfast dedication to self-reflection, personal growth, and well-being serves as imperative work in the pursuit of equity, fostering environments where every student can thrive.

FOSTERING EQUITY: CRITICAL SERVICE-LEARNING AS A TRANSFORMATIVE SOLUTION

Critical service-learning (CSL) stands as a relevant, powerful pedagogical paradigm that invites educators and students alike to delve deeply into reflection and equity work. When integrated with STEM teaching and learning, CSL offers a robust solution to address the pervasive inequities within STEM education.

The term 'service-learning' represents a type of experiential education, typically implemented within a project-based learning framework. It blends classroom theory with real-world experiences and reflection, involving partnerships and service with organizations or community members. Distinctively, CSL diverges from traditional service-learning paradigms, echoing Paulo Freire's praxis by emphasizing the fusion of reflection and action to effect transformative change (Freire, 2013).

Unlike traditional approaches, the concept of critical service-learning was first introduced by Robert Rhoads in 1997 when he explored "critical community service." Rhoads's work laid the foundation for CSL, inspiring subsequent researchers like Rice and Pollack and Rosenberger, who employed the term "critical service-learning" to describe academic service-learning experiences with a social justice orientation (Mitchell, 2008).

Tania Mitchell, a leading researcher in this field, provides guidance to educators in distinguishing between critical and traditional service-learning. She articulates, "the traditional approach emphasizes service and reflection without attention to systems of inequality, and a critical approach is unapologetic in its aim to dismantle structures of injustice" (Mitchell, 2008).

It is important for educators to understand the difference between critical service-learning and traditional service-learning to avoid the perpetuation of racism, discrimination, and white supremacy culture. Over the years, traditional service-learning has encountered criticism for its lack of reciprocity, excessive focus on student outcomes, and the potential for "service" to resemble colonization. In neglecting the historical context and power dynamics of communities, it risks transforming students into passive observers rather than informed advocates. Additionally, it inadvertently reinforces divisive "us" versus "them" narratives and inequitable hierarchies (Tinkler, Tinkler, Hausman, & Tufo-Strouse, 2014).

Critical service-learning diverges from traditional service-learning's focus on personal reflection and community service, which, though beneficial for individual and community development, may not inherently catalyze significant societal or environmental shifts. It shares the characteristics of a project-based learning (PBL) framework, reflection, and community partnerships with traditional service-learning, but it stands out due to three vital distinctions. In CSL, students are encouraged to: (1) take on active roles as agents of social change, (2) foster genuine relationships within both the classroom and the community, and (3) actively participate in the redistribution of power (Mitchell, 2008).

Essential Components of Critical Service-Learning

To unlock the potential for profound societal transformation, students must go beyond mere service and confront the root causes of the issues at hand. This paradigm shift calls for a thorough reevaluation of how educators design projects, including a critical examination of the service agencies institutions partner with. Students should be challenged to explore the underlying social problems, question the status quo, and cultivate a sense of solidarity. For example, in traditional service-learning, students might serve in a soup kitchen and reflect on their experiences. In contrast, in critical service-learning, students not only serve but also reflect on the inequitable systems contributing to food scarcity and take concrete actions to effect change in their communities.

The second focus, centered on building authentic relationships, is pivotal both in the classroom and with community partners. In the critical service-learning classroom, these genuine connections, whether they develop among students or between students and faculty, play an important role in establishing trust and serving as a model for community engagement. Achieving this necessitates several key components, including establishing classroom norms, nurturing self-awareness, engaging in community-building activities, fostering a courageous classroom community, promoting open dialogue, effective listening, and encouraging critical reflection.

When extending these principles to community partners, the process involves educators giving students the opportunity to explore and understand the community they are engaged with. Simultaneously, communities should gain insights into the students involved. It is also important for students to listen to counter-narratives from the community, fostering an environment that places a premium on mutual learning and teaching (Kokozos & Gonzalez, 2021). Additionally, an essential aspect involves fostering effective communication with community partners to pursue shared objectives and cultivate a collective understanding of the project's mission.

Moreover, it is vital to acknowledge that authenticity is not a product of a single semester; thus, it is important to establish ongoing partnerships and sustain engagement between the educational institution and the community partners. This enduring commitment results in more transformational change for both students and community partners alike.

The third element, redistributing power, is crucial, as it entails recognizing unearned privilege and unequal power dynamics among students, faculty, and community members. This recognition serves as a catalyst for fostering dialogue and reflection about these disparities. For instance, embarking on service work without talking about power and privilege may inadvertently reinforce the perception of a community in need of "fixing" (Mitchell, 2008).

In critical service-learning, recognizing power dynamics and redistributing power empower all stakeholders, especially those whose voices are often marginalized, fostering collaborative decision-making and progressing toward equitable power distribution.

CRITICAL SERVICE-LEARNING: A PEDAGOGY SUPPORTED BY RESEARCH

Motivations for Social Change in STEM

The stories of individuals like Millard, while unique in their details, underscore a common theme in their STEM journeys. Millard has an aspiration: to utilize STEM expertise to make social change. This narrative is supported by research, particularly resonating with students of color and women. In a study conducted with high school students, it was revealed that Black students' motivation for pursuing STEM careers consistently centered around a desire to help others, in contrast to their white counterparts, who cited a broader range of reasons for their career preferences (Rosenzweig & Chen, 2023). Among undergraduate STEM students in a study involving 2,697 participants, over 50 percent of underrepresented students of color considered working for social change as an essential component of their career goals, while only 37 percent of their counterparts shared this perspective (Garibay, 2015).

Furthermore, when examining the list of the most popular majors for Black students, it becomes evident that disciplines associated with public service are notably prominent. This pattern may indicate that students of color, who are more frequently exposed to negative stereotypes and injustices in American society, tend to exhibit a greater inclination toward careers focused on driving social change.

Moreover, research indicates that women tend to gravitate toward careers that offer opportunities for prosocial activities and community engagement (Rosenzweig & Chen, 2023). Additionally, there is growing evidence of students' keen interest in integrating STEM education with social justice initiatives. In a post-course survey at Carnegie Mellon University, where engineering and social justice were integrated, results showed a significant positive impact on students' engagement. Students expressed that incorporating equity concepts into engineering classes enhanced their learning by emphasizing that their designs have real impacts on people. They also expressed a desire for more direct interaction with the community and a deeper understanding of their community's history and needs to better inform their contributions (Armanios, Christian, Rooney, Mcelwee, 2021).

Relevance of Critical Service-Learning in Addressing Societal Issues

The growing student interest in effecting social change not only underscores the importance of a curriculum that fosters critical examination of inequitable systems, such as CSL, but also highlights its profound relevance within the STEM domain. For instance, following the 2014 shooting of Michael Brown, three Black teenagers developed an app for rating police interactions, illustrating technology's transformative potential in driving social change (Block, 2022).

Moreover, scientists and engineers play pivotal roles in addressing pressing issues such as greenhouse gas emissions, which have far-reaching social and environmental consequences. The widespread adoption of facial recognition technology has sparked significant concerns regarding racial bias, emphasizing the pressing need for equitable technological advancements.

Additionally, fostering diversity within STEM fields, such as recruiting BIPOC chemists specializing in formulating products for darker skin tones, can lead to more inclusive and socially responsible innovations (Block, 2022). Another example relates to the expertise of engineers in lean manufacturing that has transcended traditional boundaries. For example, at a nonprofit educational site, an engineer applied Lean manufacturing principles to their industrial kitchen, resulting in a substantial reduction in food waste, time savings, and optimized resource allocation (Pomeroy, 2023).

Empowering Students through CSL

Educators like Ron Berger have dedicated their careers to understanding the dynamics of student motivation and engagement. Through his work, he introduced the 'Hierarchy of Audience' theory, which posits that students' engagement and motivation are heightened when the audience for their learning is authentic (Berger, 2014). Berger emphasizes that the most impactful audience is one for which the product or learning is intended to serve the world. Engaging in projects aimed at affecting social or environmental change offers valuable benefits to all students (see figure 5.1).

Furthermore, research on academic mindsets, defined as psycho-social attitudes and beliefs about oneself in relation to academic work, highlights the importance of ensuring that learning remains relevant, meaningful, and connected to students' real-life experiences (Farrington, 2013). Critical service-learning effectively cultivates this mindset, providing compelling evidence that it serves as a powerful motivator for student engagement and empowers youth to cultivate a deep-seated belief in their own competence and capacity to make meaningful contributions to their communities and the broader world.

Additionally, CSL shares principles of Culturally Relevant Pedagogy (CRP), which emphasizes three core practices: sociopolitical awareness, academic excellence, and cultural competence (Hugh-Pennie et al., 2021). Cultural competence extends beyond educators' mere awareness of students' cultures; it encompasses the critical process through which students learn to honor their own heritage while navigating the culture of power, such as white,

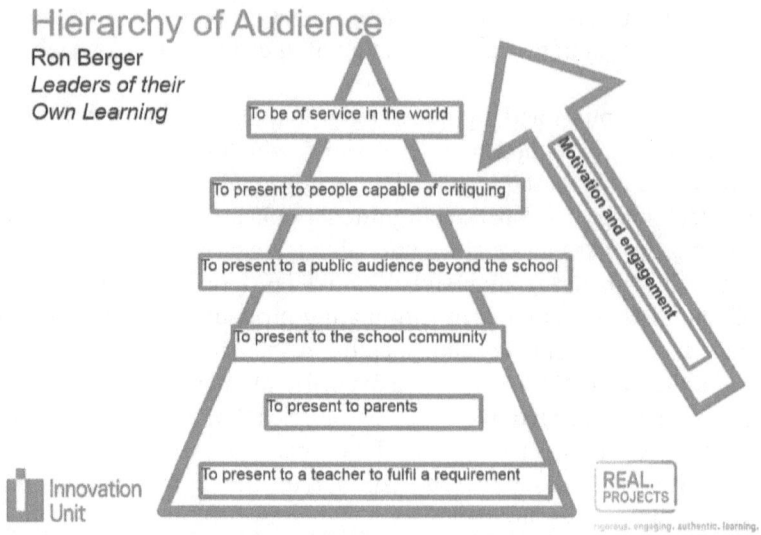

Figure 5.1 Hierarchy of Audience. *Source*: Berger, R. (2024).

middle-class culture in the United States. This aspect underscores the crucial need for a social justice curriculum that delves deeper than mere cultural exposure in classrooms—an imperative fulfilled by critical service-learning.

Example of Critical Service-Learning

While service-learning research and implementation predominantly focus on higher education, it's crucial to extend CSL to K–12 education. Essential skills like self-reflection, active listening, opinion development, solidarity, promoting social change, and nurturing authentic relationships must be explicitly taught within the K–12 system. Moreover, early exposure to relevant, equity-centered STEM pedagogy is vital for women and underrepresented students to explore the possibilities of pursuing a STEM career.

The following section highlights an example of a successful integration of STEM education and critical service-learning in a high school setting, complemented by a reflective tool provided to help students and educators discern whether their STEM projects align with traditional service-learning or critical service-learning. This tool distills the principles of both approaches into learner-centered statements, serving as a reflective guide to support alignment with the transformative goals of critical service-learning.

Traditional service-learning:

- I am aware of social and environmental problems.
- I like helping people and the environment.
- I understand my privilege and can give my time or money to help those in need.

Critical service-learning:

- I understand my *privilege* and *power* and take action to *redistribute this power.*
- I *actively listen* to the stories of community members, *amplifying marginalized voices.*
- I cultivate *meaningful connections with community partners* and engage intentionally with those impacted by injustices.
- I use *love, empathy,* and *critical thinking* to *challenge the status quo* and understand social or environmental problems.
- I *collaborate* with community members to take *action* for justice and social change.

On Our Tables

The innovative charter school network, High Tech High, is reshaping the landscape of education by authentically and profoundly advancing equity efforts. They're experts in crafting project-based learning experiences that seamlessly weave together STEM education, environmental awareness, and social justice. It's a dynamic approach that's both impactful and accessible for all students.

John Santos is a passionate Biology and Environmental Science Teacher at High Tech High Schools. In a recent project called "On Our Tables," he embarked on a mission with 12th-grade students to explore the intricate connections between our diets and our world. What sets this project apart is its seamless integration of STEM with the social sciences, transcending conventional academic boundaries to create an authentic learning experience. This journey mirrors navigating a STREAM, seamlessly blending scientific inquiry with humanities and social and environmental justice. It challenges students to contemplate the profound implications of how feeding our community impacts the environment, our health and wellness.

This captivating project begins with students reflecting on their own personal nutrition and wellness journeys. Later, students conduct interviews with agriculture experts and then individuals directly affected by food access and agriculture-related issues, synthesizing their insights into thought-provoking articles. At the project's culmination, students contribute two pages to a project publication. These pages delve into the complex web of how feeding their community, both locally and globally, influences our collective wellness.

"On Our Tables" goes beyond the classroom and reaches out into the community. In collaboration with "Barrio Botany," a San Diego-based organization dedicated to enhancing the lives of urban students through experiential garden-based learning, 12th-grade students built over 20 garden beds for donation to urban schools. These beds not only nurture plants but also impart invaluable lessons about the environment and health to many students who lack access to healthy nutrition and gardening.

John Santos eloquently encapsulates a crucial aspect of the project's effectiveness: "When you do a project that involves a community partner, the project takes on a life of its own. The engagement is deeper, the work more authentic, and the impact more profound. It's an imperative—a MUST."

"On Our Tables" sparks critical consciousness and is an example of critical service-learning in several key ways:

- It fosters an understanding of *privilege* and *power*, inspiring students to take action to redistribute power around nutrition.
- It encourages students to *actively listen* to and *amplify the voices* of community members whose stories are often overlooked.
- It encourages the development of *meaningful relationships* with community partners, such as the ongoing collaboration with Barrio Botany. This marks the second year of the partnership, further enhancing the depth and richness of the learning experience.
- It cultivates *love* and *empathy* as students strive to comprehend social and environmental injustices and question the status quo.
- It culminates in two key components: the publication of written articles and collaborative projects with Barrio Botany. Through these experiences, students learn to become *advocates* for *justice* and *social change*.

"On Our Tables" is more than just an educational project; it's a transformative experience. It equips students with the tools to become socially conscious and empathetic change-makers in their communities and beyond (Santos, 2023).

Transgress

In conclusion, the story of Millard, coupled with the exploration and exemplification of critical service-learning, underscores the pressing need to integrate STEM and social justice to create new STREAMS of learning. This integration is not just a theoretical concept; it's a dynamic, relationship-driven framework for equity and transformative change that holds immense potential for exploration and expansion.

By embracing the tenets of critical service-learning, creating environments rich in relationships, and fostering critical consciousness, STEM educators can play a pivotal role in nurturing the next generation of change agents and closing the opportunity gaps and disparities faced by women and underrepresented students in STEM. These future leaders will tackle disparities and drive progress in STEM fields, ultimately contributing to a more equitable and inclusive society.

In the spirit of bell hooks, the fusion of STEM and social justice within critical service-learning invites students to "transgress" against racial, sexual, and class boundaries, striving toward the attainment of freedom (Specia & Osman, 2015). She reminds educators that despite the limitations within the classroom, it remains a realm of possibility. It is crucial to nurture hope, even amidst seemingly insurmountable challenges. As we collectively navigate the journey to overcome boundaries and transgress, education emerges as the practice of freedom—a transformative endeavor empowering us to envision and actively create a more just and equitable world.

REFERENCES

Armanios, D. E., Christian, S.J., Rooney, A.F., Mcelwee, M. 2021). (2021). *Diversity, equity, and inclusion in civil and environmental engineering education: Social justice in a changing climate.* Presented at the 2021 ASEE Virtual Annual Conference Content Access.

Basile, V., & Azevedo, F. S. (2022, April 29). Ideology in the mirror: A loving (self) critique of our equity and social justice efforts in STEM education. *Science Education, 106*(5), 1084–1096. https://doi.org/10.1002/sce.21731.

Berger, R. (2014, January 7). *Hierarchy of Audience.* Figure, ELEducation. https://eleducation.org/resources/leaders-of-their-own-learning-chapter-6-celebrations-of-learning/

Block, D. (2022, January 9). Why STEM needs to focus on social justice. *Washington Monthly.* https://washingtonmonthly.com/2020/08/30/why-stem-needs-to-focus-on-social-justice/

Çopur-Genctürk, Y., Thacker, I., & Cimpian, J. R. (2023, April 24). Teachers' race and gender biases and the moderating effects of their beliefs and dispositions. *International Journal of STEM Education, 10*(1). https://doi.org/10.1186/s40594-023-00420-z

Costello, R. A., Salehi, S., Ballen, C., & Burkholder, E. (2023). Pathways of opportunity in STEM: Comparative investigation of degree attainment across different demographic groups at a large research institution. *International Journal of STEM Education, 10*(1). https://doi.org/10.1186/s40594-023-00436-5

Farrington, C. A. (2013). Academic mindsets as a critical component of deeper learning. *Consortium on Chicago School Research*, White Paper.

Finkel, L. (2018). Infusing social justice into the science classroom: Building a social justice movement in science education. Educational Foundations, v31 n1-2 p40-58. https://eric.ed.gov/?id=EJ1193696

Freire, P. (2013). *Pedagogy of the oppressed* (30th ed.). New York: Bloomsbury Academic.

Garibay, J. C. (2015, February 27). STEM students' social agency and views on working for social change: Are STEM disciplines developing socially and civically responsible students? *Journal of Research in Science Teaching*, 52(5), 610–632. https://doi.org/10.1002/tea.21203

Gibbs, K., Jr. (2014, September 10). *Diversity in STEM: What it is and why it matters*. Scientific American Blog Network. https://blogs.scientificamerican.com/voices/diversity-in-stem-what-it-is-and-why-it-matters/

Hugh-Pennie, A. K., Mya Hernandez, M., Uwayo, M., Johnson, G., Ross, D. (2021, September 28). Culturally relevant pedagogy and applied behavior analysis: Addressing educational disparities in PK12 schools. *Behavior Analysis in Practice*, 15(4), 1161–1169. https://doi.org/10.1007/s40617-021-00655-8.

Jin, Q. (2021, September 17). Supporting indigenous students in science and STEM education: A systematic review. *Education Sciences*, 11(9), 555. https://doi.org/10.3390/educsci11090555

Kokozos, M., & Gonzalez, M. (2021, February 15). Cultivating critical consciousness in the classroom: 10 counternarrative resources. *Getting Smart*. https://www.gettingsmart.com/2021/02/15/cultivating-critical-consciousness-in-the-classroom-ten-counternarrative-resources/

McElwee, M. (2023, September 6). Interview by author. Google Meet.

Mitchell, T. D. (2008). Traditional vs. critical service-learning: Engaging the literature to differentiate two models. *Michigan Journal of Community Service Learning*, 2008, 50–65.

O'Rourke, B. (2021, November 19). Increasing access and opportunity in STEM crucial, say experts. *Harvard Gazette*. https://news.harvard.edu/gazette/story/2021/11/increasing-access-and-opportunity-in-stem-crucial-say-experts/

Pomeroy, L. (2023, October 6). Interview by author. Google Meet.

Rosenzweig, E. Q., & Chen, X. (2023, June 9). Which STEM careers are most appealing? Examining high school students' preferences and motivational beliefs for different STEM career choices. *International Journal of STEM Education*, 10(1). https://doi.org/10.1186/s40594-023-00427-6

Santos, J. (2023, September 11). Interview by author. Google Meet.

Specia, A. Osman, A. (2015) Education as a practice of freedom: Reflections on bell hooks. https://eric.ed.gov/?id=EJ1079754

Tinkler, A. Tinkler, B., Hausman, E., Tufo-Strouse, G. (2014). Key elements of effective service learning partnerships from the perspective of community partners. *Partnerships: A Journal of Service-Learning & Civic Engagement*, 5(2), 137–152.

Chapter 6

Developing a Conceptual Framework for Early Childhood Streams Instruction

Integrating Social Emotional Competencies

Amanda Bennett

Due to the recent emphasis on science, technology, engineering, and math (STEM) careers over the past few decades, policymakers and educators are beginning to recognize the importance of helping children develop the necessary skills to succeed in these career fields and function in a world where twenty-first-century skills are highly valued (Moomaw, 2013). STEM is identified as a meta-discipline that links applications in content area disciplines to create knowledge as a whole (Johnson, 2012). STEM's educational context is commonly described as ranging from kindergarten to 12th grade; however, research on STEM education has generally emphasized upper elementary and secondary education settings (Merrill and Daugherty 2010; Moorehead and Grillo 2013). This age focus has led to limited attention and research on teaching STEM in early childhood education (ECE) settings.

An emerging body of research has revealed that early STEM experiences (preschool to third grade) can enhance children's knowledge, skills, and dispositions needed for the jobs of the future, as well as prepare students for an economy that seeks innovative solutions to complex problems (Aronin & Floyd, 2013; Chesloff, 2013; DeJarnette, 2012; New, 1999). For example, Chesloff (2013) argues that "concepts at the heart of STEM—curiosity, creativity, collaboration, and critical thinking—are in demand . . . they also happen to be innate in young children" (27). Clayton (2019) contends that delaying STEM exposure increases the chances that certain kids will never see themselves as scientists, programmers, or engineers. Thus, if STEM education does not begin until upper elementary or middle school, educators will miss numerous opportunities to take advantage of young children's natural abilities.

Despite the recent surge in implementing STEM education in schools, some researchers believed there was a component missing that could make STEM more accessible and relevant to a broader variety of students and learners (Masata, 2014). Consequently, STEM has been re-conceptualized through STEAM, where the "A" represents the arts. STEAM instruction is conceptualized as a transdisciplinary learning process that reaches across content areas and asks students to problem-solve using real-life scenarios (Quigley et al., 2017). Although STEAM education is in its early stages, especially in ECE, initial findings indicate that STEAM-based curricula increase motivation, engagement, and effective disciplinary learning in STEM areas (Kang et al., 2012).

A misconception about STEAM instruction compared to STEM is that the arts focus primarily on a finished product rather than a process of learning through thinking, planning, and creating or performing a work of art (Perignat & Katz-Buonincontro, 2019). However, arts inclusion is advantageous, as it makes the other disciplines more relevant to a broader audience of students and aids in the connection to real-world situations (Kang et al. 2012), while also incorporating a heightened focus on inclusion and equity for students who are typically underrepresented in traditional STEM fields.

Taking this idea of inclusion, equity, and integration of various disciplines a step further, researchers have introduced the concept of STREAMS (science, technology, reading/ELA, engineering, arts, mathematics, and social studies). STREAMS is another way to conceptualize and diversify STEM and STEAM so all students can feel successful, included, and represented in a meaningful and valuable way. Traditionally, STEM education emphasizes developing students' convergent thinking or problem-solving skills, whereas STEAM education focuses on divergent or creative thinking skills (Land, 2013).

Integrating convergent and divergent thinking skills can provide a more holistic representation of diversity and variety between the disciplines by introducing STREAMS education in the traditional classroom. This intersection among various disciplines can create a valuable learning experience for all students, especially students who tend to be underrepresented in traditional STEM fields. Providing students with opportunities to engage in both divergent and convergent thinking skills within educational experiences can allow them to construct meaning, increase self-motivation, and enhance learning. In the classroom, this could be demonstrated by teachers integrating components of STREAMS to allow students to decompose a complex problem using convergent thinking skills and then apply the solution to real-world scenarios using divergent thinking skills. For example, the lesson could begin by presenting students with a real-world scenario that includes a problem to be solved, thus leading to multiple possible solutions (divergent thinking). To generate the possible solutions, students must incorporate convergent thinking skills in the form of applying facts with specific answers. More specifically, a STREAMS scenario could be focused on students creating a school garden. Students would be tasked with designing, planning, and creating a garden for their school community. Within this scenario, there are many possible solutions utilizing divergent thinking that students could develop, such as garden design and aesthetics and composition of the garden (vegetables, fruits, herbs, and flowers). However, students must also use convergent thinking skills to calculate the area of the garden space and the cost of how much mulch and other materials/supplies will be needed, consideration of invasive species, climate of the area, and atmospheric conditions.

This is just one of many examples of how the inclusion of divergent and convergent thinking skills can be integrated into STREAMS educational experiences. Overall, STEAM and STREAMS education has been gaining momentum in several countries due to the emphasis on enhancing twenty-first-century skills, releasing creativity, enabling learning and practice, and promoting student-centered learning (Root-Bernstein, 2011).

INTRODUCING STREAMS IN ECE

Early childhood is the ideal developmental period to introduce STREAMS instruction for two reasons: First, there is something developmentally distinct about young children as learners that makes this stage of development appropriate for STREAMS instruction. For example, during this phase, children are primed for exploration, inquiry, innovation, curiosity, problem-solving, creative thinking, and collaboration (Chesloff, 2013; Jamil et al., 2017). Second, ECE classrooms are inherently designed to foster exploration, play-based learning, collaboration, and authentic and hands-on experiences (Christenson & James, 2015; Sharapan, 2012). Therefore, children's characteristics and abilities at this age are ideal for introducing STREAMS instruction. In addition, the many contextual strengths of early childhood classrooms present an added advantage.

At the foundation of STEAM disciplines, specifically, lie problems that need to be solved. Thus, "young children presented with a problem (science) have the ability to collaborate and work together to create a design (art) as a solution to a problem (engineering) while using available materials (technology) and strategies (math)" (DeJarnette, 2018, p. 107). This idea can be transformed to include components of reading/ELA and social studies, which allow the transition from STEAM to STREAMS to be seamlessly integrated into this new framework of understanding, thus fully encompassing the idea that the whole is greater than the sum of its parts. Including reading/ELA and social studies allows students to engage in creativity and critical thinking as part of a well-rounded curriculum.

The developmental phase of early childhood is an advantageous time to introduce students to STEAM and STREAMS instruction due to children's focus on creativity, problem-solving, and exploration (Jamil et al., 2017; Sharapan, 2012). In addition to children's love for hands-on, real-world experiences, most enjoy exploring and creating visual art, music, dance, and dramatic play (Butera et al., 2016). Further, the arts provide children with opportunities to support their creative and strategic thinking using tools and their immediate surroundings in innovative and novel ways to solve problems.

Early childhood classrooms are designed to support and foster exploration, play-based learning, authentic hands-on experiences, and collaboration among children (Bodrova, 2008; Chen & McNamee, 2011). By using centers or stations, children are afforded the opportunity to experience the unique and signature features of an ECE classroom. For example, most ECE classrooms have at least one of the following centers or areas for children to participate in: dramatic play area, water or sand table, block center, technology area, and creative arts station. While ECE classrooms' physical spaces are critical contextual elements, other aspects are also useful, such as the instruction which tends to be more unit/inquiry-based with less separation of content areas than later in school. Also, the same teacher typically works with the same students throughout the day, allowing for more fluid integration between disciplinary ideas and/or themes (Apps & MacDonald, 2012; Roskos & Neuman, 2011). Providing children with several avenues to explore, practice, and develop competencies regarding academic content and other valuable skills, such as collaboration, problem-solving, critical thinking, and emotional regulation, can be considered valuable learning opportunities.

When considering what STREAMS content, particularly the STEM disciplines, should be included in ECE, it is important to take into account students' developmental levels and the particular characteristics they possess during this phase of cognitive, social, and emotional growth. The STEM disciplines fit naturally into the way students at this age think, process information, and problem-solve. For example, young children can all be considered scientists in that they are constantly conducting experiments in their natural world, which helps them make discoveries and construct an understanding of phenomena. Children are naturally curious, inquisitive, and observant, which propels them with the drive to make sense of the world around them and positions them at a crucial stage for learning about STEAM content (DeJarnette, 2018).

In addition to children possessing scientific characteristics, young children are also adept at identifying problems and have the innate desire to solve them, which are similar to characteristics that engineers possess. Although preschool students are young and are still developing academic content knowledge, they have the capability to act as engineers and scientists by using skill sets necessary for each profession. Thus, teachers should not be hesitant to introduce complex topics related to elements of STREAMS, even to young children, as the benefits of beginning early are numerous. For example, it may spark an early interest in STEM disciplines for students, especially for populations who are typically considered underrepresented within these contexts, such as minorities and females. Also, reinforcing STREAMS content allows students the opportunity to practice valuable life skills that are crucial to becoming successful and productive citizens. Some of these skills include cooperation, divergent and convergent thinking, and compassion for others.

INTEGRATING SEC INTO STREAMS INSTRUCTION

For STREAMS instruction to be successful in any classroom, including ECE classrooms, children must possess the necessary SEC (social-emotional competencies) to carry out the activities and tasks designated by the specific STREAMS activity. These competencies are necessary if teachers and children wish to achieve STREAMS instruction's benefits, such as increased use of collaboration, inquiry-based strategies, and problem-solving skills (Sharapan, 2012). Since these skills are being developed and refined, introducing STREAMS instruction in ECE settings requires the explicit instruction of social-emotional skills.

Healthy social and emotional development in the early years lends itself to positive developmental and life outcomes in the future (Bian et al., 2017). The early childhood developmental period is an especially critical time to cultivate these instrumental social/emotional life skills, setting the stage for school readiness and success (Mann et al., 2017). For example, when beginning school, children must be able to function in groups properly, follow directions and rules, and cooperate with peers, among other things (Mathis & Bierman, 2015). Recently, an emphasis has been placed on the importance of developing children's social and emotional skills, which has led to more schools adopting social-emotional learning (SEL) programs and/or providing teachers with more professional development (PD) surrounding these skills (Greenberg et al., 2003).

Broadly, social-emotional skills and competencies are of particular importance not only for developing emotional regulation (Izard et al., 2001), cultivating healthy relationships (Rose-Krasnor, 1997), and promoting positive psychological well-being (Norona & Baker, 2014), but also for ensuring children are adequately prepared to enter formal schooling (Blair, 2002). Research shows that children with high social and emotional competence levels are better able to develop and maintain peer relationships and understand and manage emotions (Denham, 2006). Delays in emotional regulation when entering school put children at greater risk for *sustained* social and academic difficulties (McClelland et al., 2006). Furthermore, social and behavioral issues at school entry are associated concurrently and longitudinally with poor academic and social outcomes (Walker & MacPhee, 2011).

However, despite the emphasis on social-emotional skills for school success, many children lack the necessary social-emotional competence to function effectively in school (Sheridan et al., 2010). For example, 46 percent of kindergarten teachers have reported that more than half of their incoming students did not possess the foundational social and emotional competencies to succeed in school. Furthermore, 34 percent of teachers reported that more than half of their entering kindergarten students had difficulty working independently (a function of cognitive regulation), and 30 percent stated that more than half of children enter kindergarten with difficulties working cooperatively in a group (Rimm-Kaufman et al., 2000). When students have not mastered these basic social-emotional skills, they have extreme difficulty focusing and learning academic content; therefore, EC (early childhood) teachers tend to place a heavy emphasis on developing students' SEC prior to or in parallel with teaching traditional academic material.

DEFINING SEC'S COMPONENTS

Social/interpersonal skills refer to behaviors that help children interact positively and effectively with others (Jones and Bouffard 2012). Some social/interpersonal skills include recognizing and understanding social cues, effectively interpreting other's behaviors, and having positive interactions with peers and adults (McClelland et al., 2017). Specific examples of how these skills are utilized in children's daily life and/or school contexts include sharing, taking turns, cooperating, making and keeping friends, resolving conflict, and appropriately responding to various social situations.

The emotional processes component includes the skills children need to manage their emotions effectively as well as recognize others' emotions (Jones & Bouffard, 2012). These processes include skills such as emotion knowledge (the ability to recognize and label emotions accurately), emotion regulation (managing emotions and controlling how and when we express them), perspective-taking, and empathy (McClelland et al., 2017). Some examples of how these skills are demonstrated in early childhood include articulating feelings while experiencing various emotional states, understanding other's emotions, controlling inappropriate emotions, and coherently communicating instructions, thoughts, or emotions.

Finally, cognitive regulation includes attention and inhibitory control, working memory, and cognitive flexibility or set shifting (Jones & Bouffard, 2012). These cognitive regulation skills enable children to focus and switch from one task to another, listen to and remember instructions, and inhibit impulses (McClelland et al., 2017). Critical thinking and problem-solving skills are also included in this domain. Critical thinking has been a long-valued skill in society and a requirement for students and citizens to be productive and successful members of society (Allina, 2018). Critical thinking encompasses many skills that develop at different rates depending upon one's cognitive maturity and developmental level; therefore, it is critical that EC teachers consider individual students' levels of critical thinking when designing lessons and activities. Critical thinking characteristics include the ability to reason effectively, use systems thinking, make judgments and decisions, and solve problems. Some examples of these skills include active listening, following multi-step directions, planning and organizing, setting goals, and solving problems of varying complexity. Translated to ECE, this could entail developmentally appropriate levels of higher-order thinking skills such as cause-effect, similar/different, sequencing, prediction, and metacognition (Lindeman et al., 2014).

A CONCEPTUAL MODEL OF STREAMS IN ECE

Currently, a conceptual model that integrates developmentally appropriate social-emotional competence skills as an essential component of STREAMS content in ECE does not exist. This burgeoning area of research could provide ECE teachers with a framework for how to integrate foundational skills, such as SEC, with conventional academic content, evidenced in STREAMS instruction, while simultaneously addressing compulsory standards, providing assessment-driven instruction, and supporting equitable practices. To address the current gap in the literature, a new conceptual framework has been developed that integrates aspects of SEC into instructional practices and classroom environments for ECE teachers to reference and utilize in their classrooms. This framework is depicted in a process model (see figure 6.1), which will be outlined and explained below.

Both instructional practices and the classroom environment are necessary components of quality ECE settings that work together and influence one another to achieve the desired outcome, evidenced through the students' development of SEC. The SEC acquired by the learner can then be translated

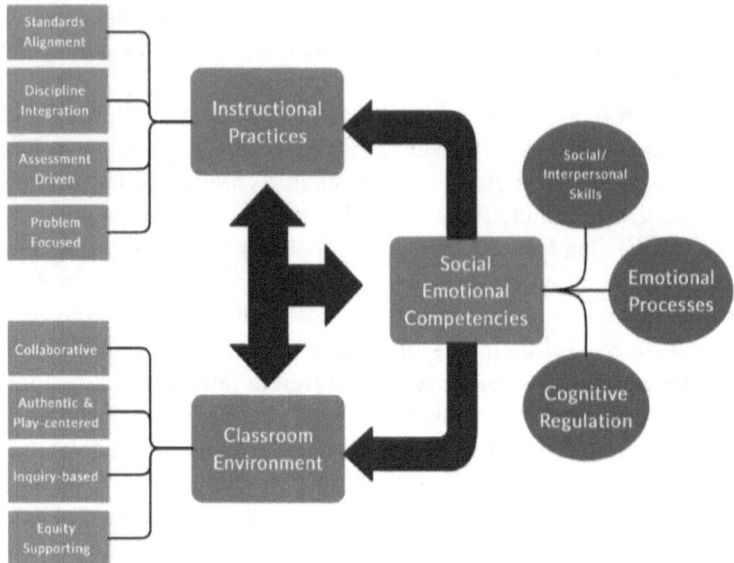

Figure 6.1 Conceptual Model: STREAMS Instruction in ECE. *Source*: A. Bennett (2024).

through participation in instructional practices and the classroom environment. This process is cyclical in that instructional practices, the classroom environment, and SEC are continuously being integrated, developed, and practiced repeatedly through students' engagement with the dimensions outlined below. Further, students' increasing proficiencies in utilizing SEC aspects will influence how they interact with the classroom environment and interpret instructional practices, hence the arrows leading back to both dimensions. This process works the other way as well, in that participation in instructional practices and the classroom environment will inform and strengthen students' SEC development.

For example, as students engage in STREAMS activities that encourage collaboration between group members to solve a common problem, their social/interpersonal skills will improve with continued practice. However, the process of learning to collaborate initially successfully needs to be explicitly guided and heavily facilitated by the teacher. Before beginning the activity, the teacher and students should discuss guidelines for how to effectively collaborate and provide specific examples or role-play what that looks like. The teacher should provide students with directive feedback and guidance throughout the process and intervene in the group as necessary. More specifically, when students work together collaboratively, all students should have a voice to openly express their ideas and feel respected by their group members (equity supporting). If the teacher notices that part of the group is not adhering to this guideline, he/she should intervene to redirect and refocus their attention on the goal they are trying to accomplish. It is crucial for teachers to be aware of and understand how these instructional practices and the classroom environment can be used to support children's SEC and vice versa.

The instructional practices domain refers to four interconnected dimensions: standards alignment, discipline integration, assessment-driven, and problem-focused. The rationale for placing these four dimensions together is that they all represent components of effective STREAMS practices and high-quality instructional strategies in ECE (Blank, 2010; Moomaw & Davis, 2010). In other words, successful STREAMS instruction implementation should include integrating multiple disciplines that are focused on real-world problems, curriculum should be designed based on the standards, and developmentally appropriate assessment strategies should be used to guide and inform current and future instruction.

The classroom environment domain refers to four interconnected dimensions: collaborative, authentic and play-centered, inquiry-based, and equity-supporting. The rationale for placing these four dimensions together is that they all exemplify fundamental characteristics within ECE classrooms (National Association for the Education of Young Children, 2009) and STREAMS instruction (Lindeman et al., 2014; Sharapan, 2012). In other words, grounding lessons and activities in authentic play-centered learning scenarios will lend itself to engaging in inquiry-based strategies that support the inclusion of equitable practices while being reinforced by an environment that encourages collaboration among students and teachers. For example, when children engage in a play-centered activity, it will naturally lead to questions and invoke curiosity (e.g., wondering why a tower of blocks fell when stacked too high or why a heavy object sank to the bottom of the water table). To solve the problem, students will need to utilize equity-supporting practices such as collaboration and communication skills.

The SEC domain has three interconnected dimensions: social/interpersonal skills, emotional competence, and cognitive regulation. The rationale for placing these three dimensions together is that these skills and competencies are the desired outcomes of participating in ECE programs that incorporate the eight dimensions embodied within instructional practices and classroom environment. In other words, children who participate in programs that foster and encourage the domains mentioned above of instructional practices and classroom environment are better equipped with the necessary skills and abilities for success in their educational experience and future careers.

In summary, this conceptual model provides teachers and practitioners with a framework for how STREAMS instruction implementation could be done in an ECE classroom. This model is unique and novel because it includes the explicit instruction of SEC and does not assume these skills already exist within each learner. Additionally, this model incorporates the essential instructional practices and classroom environment that support STREAMS instruction and embody high-quality ECE programs to achieve the desired outcomes evidenced through students' burgeoning SEC. In the future, this will help inform teachers and PD designers what successful STREAMS instruction should include and show how teachers can explicitly support the development of SEC, which is a necessary aspect of ECE for children to be successful in all aspects of their lives. However, while there are many benefits to introducing STREAMS instruction in early childhood (Chesloff, 2013; Moomaw, 2013), there are still challenges that arise when considering implementing STREAMS instruction into ECE classrooms, such as a lack of teacher knowledge and confidence surrounding STREAMS practices (Barrett, 2017), ECE teachers' beliefs about student readiness for teaching STEM disciplines in ECE (Park et al., 2017), and alignment to curriculum or state-mandated standards (Bennett & Jamil, 2022; Herro et al., 2019; Nadelson et al., 2013), to name a few. High-quality, effective PD is essential to adequately address the present challenges with implementing STEAM into ECE.

IMPLEMENTING STREAMS CONCEPTUAL MODEL IN ECE

Once teachers understand the components of the STREAMS conceptual model, it is time to implement it into their ECE classroom. This is best demonstrated through a specific example using a thematic topic. The first step when implementing this conceptual model is for teachers to decide on a topic or theme connected to their specific grade-level standards. In this example, we will focus on the theme of Earth Day through general early childhood standards related to language development and communication, mathematical thinking and expression, and scientific exploration and knowledge (South Carolina Early Learning Standards, 2014). Setting the context for the STREAMS scenario, teachers should present a scenario as a problem to be solved that is based on real-world contexts and relevant to students' personal lives.

> ***Sample STREAMS scenario:*** *Our school throws away so much trash each day and some of it can easily be recycled. Recycling is essential to help, protect, and save our environment for future generations. Recycling helps save energy, resources, and reduced waste in landfills. Also, recycling allows for the creation of new products from old products. Your job is to create a recycling center for our school. You will use materials in our classroom to design, build, and*

organize your recycling center. Then, in groups (or as a class) you will present your recycling center idea using a presentation tool of your choice.

Driving Questions*: What things at our school can be recycled? What new products could you create from recycled ones?*

All elements of STREAMS as well as SEC can be seen within this sample scenario. For example, students will need to use elements of science to figure out what items can be recycled, students will use mathematical concepts to sort and count the number of items recycled, students will learn to become productive future citizens for their communities, students will use art in creative ways to design and build their recycling centers, and use technology to research and present their projects. In order to successfully complete the project, students will need to communicate their ideas effectively, work cooperatively in groups, problem solve, use imagination and creativity, and regulate their emotions, all while learning and engaging in an academic project.

CONCLUSION

This chapter set out to propose a new model of ECE STREAMS instruction. This model draws on innovative research in STEM, STEAM, and SEC to provide a developmentally appropriate approach that integrates character skills development and academic tools more intentionally to produce lifelong learners and productive members of society. The integration of STREAMS and the explicit instruction on students' SEC in the model proposed represents a natural alignment with ECE's fundamentals, in that students who engage in STREAMS education must be capable of successfully managing emotional processes, employing social/interpersonal skills, and utilizing appropriate cognitive skills (e.g., critical thinking, creativity, and attention). Students who have not mastered these SEC will struggle to be successful in any type of instructional practice; therefore, SEC should be viewed as an integral and foundational part of STREAMS education, rather than a separate entity that is unrelated or disconnected from traditional STREAMS practices.

Similarly, while SEL programs are valuable, students in the early childhood developmental stage have difficulty translating the lessons and skills acquired to other aspects of their educational experience (Garner et al., 2018). It is far more advantageous for ECE settings to establish an integrated working partnership between developing students' SEC and engaging in STREAMS instruction. The proposed conceptual model illustrates that relationship and takes into consideration the role and importance of content, process, and learner skills in both STREAMS and SEC. For this newly proposed conceptual framework linking STREAMS and the explicit instruction of SEC to be effective, ECE teachers must receive high-quality PD opportunities that explicate the relationship between STREAMS and SEC and provide strategies on how to successfully incorporate these two entities into the curriculum and classroom. Using extant literature to propose the integration of STREAMS and SEC, a valuable next step to developing high-quality PD would be to first investigate how teachers view these two types of education and their relationship in the early childhood years. A more responsive PD addressing an existing need could be developed by better understanding teachers' views and perceptions of each of these entities.

REFERENCES

Allina, B. (2018). The Development of STEAM educational policy to promote student creativity and social empowerment. *Arts Education Policy Review*, *119*(2), 77–87.

Apps, L., & MacDonald, M. (2012). Classroom aesthetics in early childhood education. *Journal of Education and Learning*, *1*(1), 49–59.

Aronin, S., & Floyd, K. K. (2013). Using an iPad in inclusive preschool classrooms to introduce STEM concepts. *Teaching Exceptional Children*, *45*(4), 34–39.

Barrett, D. (2017). STEAM framework feasibility study. http://child360.org/wpcontent/uploads/2019/08/LAUP_CA_FULLREPORT_STEAMFFS_rev20170619.pdf

Bennett, A., & Jamil, F. (2022). Kindergarten teacher responses to a contextualized professional development workshop on STEAM teaching. *International Journal of Teacher Education and Professional Development (IJTEPD)*, *5*(1), 1–15.

Bian, X., Xie, H., Squires, J., & Chen, C-Y. (2017). Adapting a parent-completed, socioemotional questionnaire in China: The ages & stages questionnaires: Social-emotional. *Infant Mental Health Journal*, *38*(2), 258–266.

Blair, C. (2002). School readiness: Integrating cognition and emotion in a neurobiological conceptualization of children's functioning at school entry. *American Psychologist*, *57*(2), 111–127.

Blank, J. (2010). Early childhood teacher education: Historical themes and contemporary issues. *Journal of Early Childhood Teacher Education*, *31*(4), 391–405.

Bodrova, E. (2008). Make-believe play versus academic skills: A Vygotskian approach to today's dilemma of early childhood education. *European Early Childhood Education Research Journal*, *16*(3), 357–369.

Borko, H. (2004). Professional development and teacher learning: Mapping the terrain. *Educational Researcher*, *33*(8), 3–15.

Butera, G., Horn, E. M., Palmer, S. B., Friesen, A., & Lieber, J. (2016). Understanding Science, Technology, Engineering, Arts, and Mathematics (STEAM), within early childhood special education. In B. Reichow, B. A. Boyd, E. E. Barton, & S.

L. Odom (Eds.), *Handbook of early childhood special education* (pp. 143–161). Springer International Publishing.

Chen, J.-Q., & McNamee, G. D. (2011). Positive approaches to learning in the context of preschool classroom activities. *Early Childhood Education Journal, 39*(1), 71–78.

Chesloff, J. D. (2013). Why STEM education must start in early childhood. *Education Week, 32*(23), 27–32.

Christenson, L. A., & James, J. (2015). Building bridges to understanding in a preschool classroom: A morning in the block center. *YC: Young Children, 70*(1), 26–31.

Clayton, V. (2019). STEM ming the myths: Students see more success in K–12 when they tackle STEM concepts in early grades. *District Administration, 55*(4), 20–23.

DeJarnette, N. K. (2018). Implementing STEAM in the early childhood classroom. *European Journal of STEM Education, 3*(3), 1–9.

DeJarnette, N. K. (2012). America's children: Providing early exposure to STEM (Science, Technology, Engineering and Math) initiatives. *Education, 133*(1), 77–84.

Denham, S. A. (2006). Social-emotional competence as support for school readiness: What is it and how do we assess it? *Early Education and Development, 17*(1), 57–89.

Dickinson, D. K., & Brady, J. P. (2006). Toward effective support for language and literacy through professional development. In M. J. Zaslow & I. Martinez-Beck (Eds.), *Critical issues in early childhood professional development* (pp. 141–170). Brookes Publishing Company.

Garet, M. S., Porter, A. C., Desimone, L., Birman, B. F., & Yoon, K. S. (2001). What makes professional development effective? Results from a National sample of teachers. *American Educational Research Journal, 38*(4), 915–945.

Garner, P. W., Gabitova, N., Gupta, A., & Wood, T. (2018). Innovations in science education: Infusing social emotional principles into early STEM learning. *Cultural Studies of Science Education, 13*(4), 889–903.

Greenberg, M. T., Weissberg, R. P., O'Brien, M. U., Zins, J. E., Fredericks, L., Resnik, H., & Elias, M. J. (2003). Enhancing school-based prevention and youth development through coordinated social, emotional, and academic learning. *American Psychologist, 58*(6–7), 466–474.

Herro, D., Quigley, C., & Cian, H. (2019). The challenges of STEAM instruction: Lessons from the field. *Action in Teacher Education (Routledge), 41*(2), 172–190.

Izard, C., Fine, S., Schultz, D., Mostow, A., Ackerman, B., & Youngstrom, E. (2001). Emotion knowledge as a predictor of social behavior and academic competence in children at risk. *Psychological Science, 12*(1), 18–23.

Jamil, F. M., Linder, S. M., & Stegelin, D. A. (2018). Early childhood teacher beliefs about STEAM education after a professional development conference. *Early Childhood Education Journal, 46*(4), 409–417.

Johnson, C. C. (2012). Implementation of STEM education policy: Challenges, progress, and lessons learned. *School Science & Mathematics, 112*(1), 45–55.

Jones, S. M., & Bouffard, S. M. (2012). Social and emotional learning in schools from programs to strategies. *Social Policy Report, 26*(4), 3–22.

Kang, M., Park, Y., Kim, J., & Kim, Y. (2012). *Learning outcomes of the teacher training program for STEAM education.* 2012 International Conference for Media in Education, Beijing, China.

Land, M. H. (2013). Full STEAM ahead: The benefits of integrating the arts into STEM. *Procedia Computer Science, 20,* 547–552.

Lindeman, K. W., Jabot, M., & Berkley, M. T. (2013). The role of STEM (Or STEAM) in the early childhood setting. In L. E. Cohen & S. Waite-Stupiansky (Eds.), *Learning across the early childhood curriculum* (pp. 95–114). Emerald Group Publishing Limited.

Mann, T. D., Hund, A. M., Hesson-McInnis, M., & Roman, Z. J. (2017). Pathways to school readiness: Executive functioning predicts academic and social-emotional aspects of school readiness: Executive functioning and school readiness. *Mind, Brain and Education, 11*(1), 21–31.

Masata, D. (2014). Understanding the STEM skills gap. *STEM Education News.* http://www.educationandcareernews.com

Mathis, E. T. B., & Bierman, K. L. (2015). Dimensions of parenting associated with child prekindergarten emotion regulation and attention control in low-income families: Emotion regulation and attention control. *Social Development, 24*(3), 601–620.

McClelland, M. M., Acock, A. C., & Morrison, F. J. (2006). The impact of kindergarten learning-related skills on academic trajectories at the end of elementary school. *Early Childhood Research Quarterly, 21*(4), 471–490.

Merrill, C., & Daugherty, J. (2010). STEM education and leadership: A mathematics and science partnership approach. *Journal of Technology Education, 21*(2), 21–34.

Moomaw, S. (2013). *Teaching STEM in the early years activities for integrating science, technology, engineering, and mathematics.* Redleaf Press.

Moomaw, S., & Davis, J. A. (2010). STEM comes to preschool. *YC Young Children, 65*(5), 12–18.

Moorehead, T., & Grillo, K. (2013). Celebrating the reality of inclusive STEM education: Co-teaching in science and mathematics. *Teaching Exceptional Children, 45*(4), 50–57. https://doi.org/10.1177/004005991304500406

Nadelson, L., S., Callahan, J., Pyke, P., Hay, A., Dance, M., & Pfiester, J. (2013). Teacher STEM perception and preparation: Inquiry-based STEM professional development for elementary teachers. *The Journal of Educational Research, 106*(2), 157–168. https://doi.org/10.1080/00220671.2012.667014

National Association for the Education of Young Children. (2009). *Developmentally appropriate practice in early childhood programs serving children from birth to age 8.* National Association for the Education of Young Children. https://www.naeyc.org/sites/default/files/globally-shared/downloads/PDFs/resources/position-statements/PSDAP.pdf

New, R. S. (1999). Playing fair and square: Issues of equity in preschool mathematics, science, and technology. In *Dialogue on early childhood science, mathematics, and technology education* (pp. 138–156). American Association for the Advancement of Science.

Norona, A. N., & Baker, B. L. (2014). The transactional relationship between parenting and emotion regulation in children with or without developmental delays. *Research in Developmental Disabilities, 35*(12), 3209–3216. https://doi.org/10.1016/j.ridd.2014.07.048

Park, M.-H., Dimitrov, D. M., Patterson, L. G., & Park, D.-Y. (2017). Early childhood teachers' beliefs about readiness for teaching science, technology, engineering, and mathematics. *Journal of Early Childhood Research: ECR, 15*(3), 275–291. https://doi.org/10.1177/1476718X15614040

Perignat, E., & Katz-Buonincontro, J. (2019). STEAM in practice and research: An integrative literature review. *Thinking Skills and Creativity, 31,* 31–43. https://doi.org/10.1016/j.tsc.2018.10.002.

Quigley, C. F., Herro, D., & Jamil, F. M. (2017). Developing a conceptual model of STEAM teaching practices: Developing a conceptual model. *School Science and Mathematics, 117*(1–2), 1–12. https://doi.org/10.1111/ssm.12201

Rimm-Kaufman, S., Pianta, R. C., & Cox, M. J. (2000). Teachers' judgments of problems in the transition to kindergarten. *Early Childhood Research Quarterly, 15*(2), 147–166. https://doi.org/10.1016/S0885-2006(00)00049-1

Root-Bernstein, M. (2011). From STEM to STEAM to STREAM: Writing as an essential component of science education. *Psychology Today*, last modified March 16. Retrieved September 14, 2023, from https://www.psychologytoday.com/intl/blog/imagine/201103/stem-steam-stream-writing-essential-component-science-education

Rose-Krasnor, L. (1997). The nature of social competence: A theoretical review. *Social Development (Oxford, England), 6*(1), 111–135. https://doi.org/10.1111/1467 9507.00029

Roskos, K., and Neuman, S. B. (2011). The classroom environment: First, last, and always. *The Reading Teacher, 65*(2), 110–114. https://doi.org/10.1002/TRTR.01021

Schachter, R. E., Gerde, H. K., & Hatton-Bowers, H. (2019). Guidelines for selecting professional development for early childhood teachers. *Early Childhood Education Journal, 47*(4), 395–408. https://doi.org/10.1007/s10643 019 00942 8

Sharapan, H. (2012). From STEM to STEAM: How early childhood educators can apply Fred Rogers' approach. *YC Young Children, 67*(1), 36–40.

Sheridan, S. M., Knoche, L. L., Edwards, C. P., Bovaird, J. A., & Kupzyk, K. A. (2010). Parent engagement and school readiness: Effects of the getting ready intervention on preschool children's social-emotional competencies. *Early Education and Development, 21*(1), 125–156. https://doi.org/10.1080/10409280902783517

Standards. (2014). South Carolina department of education. https://ed.sc.gov/instruction/early-learning-and-literacy/early-learning/standards/

Walker, A. K., & MacPhee, D. (2011). How home gets to school: Parental control strategies predict children's school readiness. *Early Childhood Research Quarterly, 26*(3), 355–364. https://doi.org/10.1016/j.ecresq.2011.02.001

Chapter 7

Building a Diverse Stream of Engineers through Teacher Research Experiences in Engineering

The IDEA-BioE Project

Meagan M. Patterson, Prajnaparamita Dhar, and Douglas Huffman

Interest in engineering among pre-college students, especially girls and students from racial and ethnic minority groups, is often low, in part due to students' lack of understanding of what engineering is. This suggests the need for rethinking how the field might introduce engineering concepts at an early age since many students make decisions about STEM early in their schooling. However, simply exposing students to engineering design principles is not sufficient to create interest in engineering careers among a broad range of students; interventions must also address social and psychological barriers (Wang & Degol, 2017). The metaphor of streams and the concept of integrating disciplines through STREAMS (Science, Technology, Reading/Language Arts, Engineering, Arts, Mathematics, and Social Studies) both help us to think about all the barriers some students face to enter STEM fields and cultural shifts that we may need to draw more students into STEM, and specifically engineering.

In this chapter, we identify major challenges to promoting interest in engineering and engineering careers among K–12 students. First, teachers may not be aware of the scope and specifics of the engineering field. Engineering draws on a range of disciplines, including mathematics, physics, chemistry, and biology, and incorporates a broad range of subfields, such as chemical, electrical, environmental, mechanical, and biomedical engineering. The social sciences disciplines are also a component of engineering that are often ignored when students engage in engineering. The creativity and design process used in the Arts translates to engineering, the ways in which engineers communicate their work rely on reading and language arts, and engineers' work affects societies from a social studies perspective. In addition, although engineering draws on science concepts, the fundamental approaches used in engineering are meaningfully different from those used in the sciences (Bybee, 2011) and thus teachers who are primarily trained in science pedagogy may struggle to incorporate and explain engineering concepts within science courses. Science teachers may not consider how to integrate other disciplines into science courses in meaningful ways.

Students may also be disinclined to engage with engineering for many reasons. Students may be unaware of what engineers do (Owen, 2011). Students may struggle to see themselves in the engineering field due to racial and gender stereotypes and a lack of role models similar to themselves within engineering specifically and STEM more broadly (Del Toro & Wang, 2023; Shin et al., 2016). In addition, perceptions of engineering as a field that primarily involves working alone, requires brilliance to be successful, and is done by those who are nerdy or socially awkward (Cheryan et al., 2015; Meyer et al., 2015; Miller et al., 2006) may negatively impact student interest in engineering. Students with strengths in disciplines such as the arts, reading/language arts, and social sciences may not know how those strengths can be an advantage to draw on when engaging in the STEM fields, particularly in engineering.

Our project's purpose was to address these barriers through a National Science Foundation-supported research experience program for teachers. The first part of our program involved an immersive six-week summer experience that aimed to: (a) educate pre-service and practicing teachers about current research in bioengineering, (b) help teachers translate their new knowledge of bioengineering into engineering design (ED) modules appropriate for classroom implementation, and (c) incorporate relevant constructs from educational psychology into ED modules. Following this research experience and module development, teachers implemented the modules in a classroom setting.

BACKGROUND

Problem Statement

Although the numbers of women and racial/ethnic minorities (i.e., African Americans, Latinos, and Native Americans) in engineering have increased over time, members of these groups remain underrepresented in engineering relative to the overall workforce. Interest in engineering and other STEM careers is generally established by the time students get to college

(Maltese & Tai, 2011). Therefore, it is important to create opportunities for exposure to engineering in K–12 education to build a "stream" of students interested in engineering careers prior to college. Engaging in project-based STEM learning during the middle school years encourages consideration of STEM careers (Christentsen et al., 2015). However, students of color may have less access to such programs than White students (Thomas & Larwin, 2023).

We identify major challenges to creating these engineering interest streams among K–12 students. First, teachers may lack awareness of key elements of engineering design, engineering careers, and current innovations in engineering. Similarly, students may lack knowledge about engineering and engineering careers. Stereotypes regarding race, gender, and STEM, as well as stereotypes of engineers, may also reduce interest in and sense of "fit" with engineering for many students, especially those from groups currently underrepresented in the engineering field.

Teacher Factors

As is common among adults (Gibson & Hutton, 2017), many teachers may be largely unaware of the nature of engineering and engineering careers, including the types of questions and problems that engineers address and the overall breadth of the field. In addition, engineering is rapidly changing as new subfields and concepts emerge. State-of-the-art engineering today relies upon the integration of multiple disciplines. However, most teachers are likely unaware of many innovative aspects of contemporary engineering design, such as the use of bioengineered products to treat injuries and diseases.

In addition, engineering design practices (EDP) differ meaningfully from science inquiry practices (SIP), and because STEM education in the United States tends to focus on mathematics and science, teachers may lack expertise or support for teaching engineering design. Teachers differ widely in what they consider engineering to be and how they implement engineering concepts in their classes (Katehi et al., 2009). Consistent with these challenges, a study of middle school teachers showed that while they were able to describe aspects of integrated STEM education, when asked to draw their own conceptions, "nontraditional subjects like engineering and technology were sometimes represented in subordinate ways to traditional subjects like science and mathematics" (Kloser et al., 2018, p. 335). These findings suggest that teachers need additional support to integrate engineering design practices into their classes, particularly those that bridge traditional domains (Lie et al., 2019; Gardner et al., 2019).

Student Factors

Research indicates that students' interest in engineering and other STEM fields decreases by middle school, and that this decline is often sharper among girls and students of color relative to other students (Herbert & Stipek, 2005; Master, 2021; Wiebe et al., 2013). One possible explanation is the perception among students that women are not suited to science and engineering fields (i.e., gender stereotyping; Master, 2021). Although less extensively studied than gender stereotypes, racial stereotypes likely also play a role in student disengagement with STEM and the underrepresentation of people of color in the engineering field (Beasley & Fischer, 2012; Del Toro & Wang, 2023). Stereotypes may contribute to a lack of self-efficacy and confidence in engineering-related skills. Low levels of confidence and self-efficacy in STEM may partially explain the lack of representation of women and people of color in the engineering workforce (Lie et al., 2019; Master, 2021; Thomas & Larwin, 2023).

USING PSYCHOLOGY-INFORMED APPROACHES TO ADDRESS BARRIERS TO STUDENT ENGAGEMENT

The IDEA-BioE project was designed to incorporate a number of concepts from educational psychology to promote student motivation and engagement. These include addressing stereotypes, promoting self-efficacy, building a growth mindset, incorporating role models, considering occupational values, and reflecting on language. The goal is to create a new way to help underrepresented students engage in engineering and create a space that truly belongs to all students who do not always fit the norm.

Addressing Stereotypes

Stereotypes promote the view that engineering is a field that is not for everyone and may contribute to the underrepresentation of women and people of color in engineering fields. Stereotypes are tied to students' own interest and self-perceptions in STEM (Cvencek et al., 2011; Master et al., 2021; Plante et al., 2013; Starr & Simpkins, 2021). Parents and teachers may reinforce these stereotypes, either intentionally or inadvertently (Chestnut et al., 2021; Master, 2021). Students pay attention to signals and messages that are relevant to stereotypes, such as differential treatment by teachers (Rubie-Davies & Peterson, 2010; Leaper & Brown, 2014). Along with gender and racial stereotypes, stereotypes of scientists and engineers (such as "nerd," "genius," or "brilliant") can influence students' motivation to engage with STEM (Bian et al., 2017; Master & Meltzoff, 2016).

However, it is important to acknowledge that not all students endorse stereotypes, and some openly reject or resist them (Asamani et al., 2022; Way et al., 2013). There are various approaches teachers can take to address and dismantle stereotypes in the classroom. Interventions to address stereotypes in the STEM arena take a number of forms, including emphasizing compatibility between STEM and traditionally feminine skills or preferences and explicitly challenging gender stereotypes (Liben & Coyle, 2014). Such interventions try to address various mechanisms of the STEM gender gap, including self-efficacy, identity, and anxiety.

Promoting Self-Efficacy

STEM self-efficacy (i.e., belief in one's own ability and competence in STEM domains) is an important predictor of student interest in STEM and intentions to persist in STEM (Brown et al., 2016; Lie et al., 2019; Master, 2021). It is also important to acknowledge that gender and racial differences in self-efficacy may vary across STEM disciplines; for example, Wilson et al. (2015) found gender gaps in self-efficacy in some STEM disciplines (including engineering) but not others. Many factors contribute to students' self-efficacy, including previous personal experiences, messages from authority figures, and observing models (i.e., vicarious learning; Bandura, 1997).

Mastery experiences (i.e., successful completion of relevant tasks) are critical to developing a sense of self-efficacy for STEM (Britner & Pajares, 2006; Kiran & Sungur, 2012). Thus, opportunities to engage with engineering design tasks in their classrooms should promote students' self-efficacy for engineering. Teachers' self-efficacy for science can also serve to promote self-efficacy in their students through modeling (Opperman et al., 2019); this may be especially true for girls since female science teachers can serve as important models for female students (Master et al., 2014). Providing information about women and people of color in science and engineering who can serve as role models (as discussed below) can also promote self-efficacy.

Building Growth Mindset

Growth mindset encompasses a set of beliefs regarding intelligence, ability, and success. Students with growth mindsets believe that intelligence and ability can improve with time and effort, rather than being fixed (Dweck, 2006; Stohlmann, 2022). Growth mindsets can promote positive expectations, reduce the impact of stereotypes, and promote academic achievement (Degol et al., 2018; Porter et al., 2022). Growth mindsets help students to persist and maintain confidence when they are not immediately successful with a task or domain (Dweck, 2006; Yeager & Dweck, 2020).

Teachers and other adults can promote growth mindsets by focusing on effort and strategy choice (rather than ability) in the feedback that they give to students, framing struggle as a natural and expected part of the learning process, and encouraging students to take on challenging tasks (Dweck, 2006; Haimovitz & Dweck, 2017). Even brief growth mindset interventions can be useful for reducing stereotypes regarding STEM fields (Law et al., 2021), but more extensive interventions that are consistently incorporated into classroom environments may have greater benefits (Haimovitz & Dweck, 2017).

Incorporating Role Models

Many students lack engineering role models and opportunities to learn about the engineering profession and the engineering design process within their own communities. Children's occupational interests are influenced by who they see performing specific occupations, including representation by gender and race (Hayes et al., 2018; Hughes & Bigler, 2008). Thus, presenting a variety of examples of role models within the engineering field is important for promoting a diverse stream of students interested in engineering careers. Simple presentation of women in engineering as role models may not be sufficient to increase girls' interest, however. Portrayals of female engineers may inadvertently backfire when these presentations reinforce gendered stereotypes of engineering (e.g., focusing on masculine-typed interests such as comic books or video games; Cheryan et al., 2013a, 2015). In contrast, showing students that engineers do not all fit traditional stereotypes may help to increase interest in the field (Cheryan et al., 2013b). Moreover, taking this notion one step further can lead to dismantling stereotypes in such a way that women can create engineering spaces that are truly designed by and for women.

To promote student engagement, teachers should incorporate examples of counter-stereotypic models with a range of personalities and interests (Cheryan et al., 2013b; Stout et al., 2011). Presenting information about scientists and engineers from a variety of identities, as well as directly addressing the biases and challenges that individuals from underrepresented groups face, can help to promote student motivation and engagement and encourage relevant career aspirations (Pahlke et al., 2010; Shin et al., 2016; Weisgram & Bigler, 2007).

Considering Occupational Values

Goals and values shape students' occupational aspirations from early in life (Hayes et al., 2018; Weisgram & Bigler, 2006). In addition, experiences and environments that reinforce one's values promote a sense of purpose, integrity, and belonging. Girls and women tend to place more weight on communal goals and values (e.g., working collaboratively, helping others) than boys and men (Diekman et al., 2010; Weisgram et al., 2011). This emphasis on communal values may contribute to the underrepresentation of women and people of color in engineering and other STEM fields, since these fields are often viewed as incongruent with communal values (Diekman et al., 2010; Gibson & Hutton, 2017).

Emphasizing the altruistic and communal aspects of STEM and STEM careers (such as helping others and working collaboratively) can help to promote interest in STEM and STEM self-efficacy for girls, women, and students of color (Brown et al., 2015; Dasgupta et al., 2022; Diekman et al., 2010; Falco & Summers, 2019; Smith et al., 2014). Interventions that address students' values can have a greater impact on their career aspirations than interventions that merely expose students to career-related content or promote self-efficacy (Weisgram & Bigler, 2006).

Reflecting on Language

There are a variety of ways of talking about science, including action-focused (e.g., "science is a way of discovering new things") and identity-focused (e.g., "scientists work hard and solve problems") approaches (Wang et al., 2022). Promoting the idea of science as something that a person does (i.e., an action-focused approach), rather than a scientist as a kind of person that someone is, can increase students' inclusive beliefs about science and personal interest in science (Rhodes et al., 2019; Wang et al., 2022). In contrast, identity-focused descriptions of science can create the idea that scientists are a distinct kind of person, and a given individual might or might not be that kind of person (Lei et al., 2019; Rhodes et al., 2019). In addition, well-intentioned messages about inclusion in STEM, such as saying that girls are just as good as boys at mathematics, can backfire by treating one group as the default or reference point (Chestnut et al., 2021).

Teachers can help promote student engagement with engineering by using action-focused language (e.g., "today we are going to practice engineering," "engineering is a way to use science to solve problems") that promotes a view of engineering as an action in which a wide variety of people can engage. In addition, teachers should avoid making group comparisons or associating identity categories (such as gender and race) with particular academic domains or career paths, even if such comparisons are intended to be positive (Bigler et al., 2017; Master & Meltzoff, 2020).

PROJECT DESCRIPTION

IDEA-BioE is a four-year NSF-funded project that involves several components: on-campus experiential opportunities for pre-service and practicing teachers, creation of classroom modules, and implementation of created modules in the classroom. (This project was funded in part by the National Science Foundation, NSF, EEC Award #2055716.) The experiential learning portion of the program involves a six-week summer experience at the University of Kansas, hosted primarily in the School of Engineering, with faculty leaders from various disciplines including chemistry, chemical engineering, mechanical engineering, educational psychology, and STEM education.

Project participants include both pre-service and practicing teachers. Pre-service teachers are recruited from teacher education programs at the University of Kansas. Practicing teachers are recruited from local school districts, with a focus on schools that have high percentages of students of color and/or students eligible for free or reduced-price lunch. This enables the program to have the greatest impact on students from underrepresented populations in the engineering field.

Participants are matched with research lab experiences from various disciplines including chemistry, chemical engineering, and mechanical engineering; all partner labs are conducting research relevant to bioengineering. Bioengineering is a field of engineering with a focus on solving real-world problems, which can be particularly appealing to underrepresented students (Thoman et al., 2015). Participants work in their relevant labs for three to four days per week during the summer program. Teachers generally work in pairs in these research labs to allow opportunities for near-peer interactions. Faculty members, graduate students, and postdoctoral fellows serve as research mentors and work with the teachers to train them on specific research equipment, research protocols, and lab safety protocols.

Typically, teachers work on a relatively small project that is a part of the lab's primary ongoing research projects and has research goals that can be met within the time limits of the summer experience. For example, in one project, teachers learned about how chemical engineers are contributing to the stability of protein-based drugs to treat various diseases by designing experimental systems that can help predict a protein-based drug's stability during biomanufacturing and storage. In another project, teachers worked with nanocellulose, a nanomaterial derived from plant fibers, as a potential option to replace petroleum-based adhesives. Teachers experienced how engineers are using biotechnology to derive molecules that can replace petroleum-based products, and thus establish more sustainable engineering approaches.

While teachers are working on their assigned research projects, they also work with faculty mentors to translate these complex research problems into learning modules appropriate for secondary school learners. As part of this process, teachers attend module development workshops that address engineering design, next-generation science standards, and best practices for lesson design (e.g., incorporating formative assessment). The workshops also discuss educational psychology concepts, including growth mindset, role models, occupational values, STEM anxiety, and language impact.

Pre-service and practicing teachers work together in pairs to create classroom modules based on their lab experiences. Topics of the created modules from the program include water filtration, surfactants, myelination and its impacts on health, and the use of hydrogels in glue. After the program's completion, teachers implement the modules in classrooms. The modules are also shared with other teachers through a variety of dissemination outlets.

PROGRAM OUTCOMES

We are fortunate to be located so close to the historic site of the *Brown v. Board of Education* court case that pushed the United States toward desegregation of our schools. The Brown case was based on the view that separate education is not equal education. However, more than 70 years later we still have not realized the vision of Brown. Many schools today are still highly segregated, and spaces that welcome underrepresented students into STEM are rare. Feedback from participants in the BioE project indicates that the project has impacted participating teachers in a variety of ways. Participating teachers completed surveys of their knowledge and attitudes regarding engineering, including questions about what engineering is, what engineers do on a typical day, and who they picture when they picture an engineer. Future research will explore the program's impact on the students of participating teachers after module implementation.

Descriptions of what engineering is and who participants pictured when picturing an engineer changed meaningfully from pre- to post-test. Prior to the IDEA-BioE program, teachers' descriptions of engineers typically focused on characteristics such as intelligence and being good at mathematics and science. Post-intervention responses included more discussion of skills beyond mathematics and science, such as communication, collaboration, and creativity. Responses also indicated broader views of engineering, including: "I can now picture almost anyone as an engineer"; "an engineer is not who someone is, it's what they do"; and "anyone can be an engineer."

Perceptions of engineering as a career were more positive at post-test than at pre-test in a manner consistent with the program's focus. For example, endorsement of the statement that engineering is a field that helps people increased from 4.69 to 5.00 (on a one-to-five scale) from pre- to post-test. In addition, endorsement of the statement that engineering is a competitive (rather than collaborative) field decreased from 3.54 to 2.67 from pre- to post-test. Thus, participating teachers were more aware of communal aspects of engineering after completing the program.

Now the goal is to take these positive indications from the BioE Project and move toward creating spaces and experiences for women and other underrepresented students that can truly welcome all students. We know that female and underrepresented students do not all follow the same path as traditional STEM students. We look forward to the time when all students have access to and opportunity for diverse, equitable, inclusive environments that help all students feel like they belong in engineering.

REFERENCES

Asamani, G. A., Adjapong, E. S., & Emdin, C. (2022). Exploring how a hip-hop based science program afforded Black/Brown girls the space to resist against Black/Brown negative stereotypes in STEM. *Journal of Urban Learning, Teaching, and Research, 16*(2), n2.

Bandura, A. (1997). *Self-efficacy: The exercise of control.* W. H. Freeman.

Beasley, M. A., & Fischer, M. J. (2012). Why they leave: The impact of stereotype threat on the attrition of women and minorities from science, math and engineering majors. *Social Psychology of Education, 15*, 427–448. https://doi.org/10.1007/s11218-012-9185-3

Bian, L., Leslie, S. J., & Cimpian, A. (2017). Gender stereotypes about intellectual ability emerge early and influence children's interests. *Science, 355*(6323), 389–391. https://www.science.org/doi/full/10.1126/science.aah6524

Bigler, R. S., Hayes, A. R., & Patterson, M. M. (2017). Social striving: Social group membership and children's motivations and competencies. In A. J. Elliot, C. S. Dweck, & D. S. Yeager (Eds.), *Handbook of competence and motivation: Theory and application* (2nd ed., pp. 547–565). Guilford.

Britner, S. L. & Pajares, F. (2006). Sources of science self-efficacy beliefs of middle school students. *Journal of Research in Science Teaching, 43*(5), 485–499. https://doi.org/10.1002/tea.20131

Brooks, J., (2020). Why should I care about diversity in engineering? *PE Magazine.* https://www.nspe.org/resources/pe-magazine/july-2020/why-should-i-care-about-diversity-engineering

Brown, E. R., Thoman, D. B., Smith, J. L., & Diekman, A. B. (2015). Closing the communal gap: The importance of communal affordances in science career motivation. *Journal of Applied Social Psychology, 45*(12), 662–673. https://doi.org/10.1111/jasp.12327

Brown, P. L., Concannon, J. P., Marx, D., Donaldson, C., & Black, A. (2016). An examination of middle school students' STEM self-efficacy, interests and perceptions. *Journal of STEM Education, 17*(3), 27–38. https://www.jstem.org/jstem/index.php/JSTEM/article/view/2137/1784

Bybee, R. W. (2011). Scientific and engineering practices in K–12 classrooms. *Science Teacher, 78*(9), 34–40.

Cagle, N.L., Caldwell, L., & Garcia, R. (2018). K–12 diversity pathway programs in the E-STEM fields: A review of existing programs and summary of unmet needs. *Journal of STEM Education, 19*(4), 12–18. https://www.jstem.org/jstem/index.php/JSTEM/article/view/2324/2084

Cheryan, S., Drury, B. J., & Vichayapai, M. (2013a). Enduring influence of stereotypical computer science role models on women's academic aspirations. *Psychology of Women Quarterly, 37*(1), 72–79. https://doi.org/10.1177/0361684312459328

Cheryan, S., Plaut, V. C., Handron, C., & Hudson, L., (2013b) The stereotypical computer scientist: Gendered media representations as a barrier to inclusion for women. *Sex Roles, 69*(1), 58–71. https://doi.org/10.1007/s11199-013-0296-x

Cheryan, S., Master, A., & Meltzoff, A. N. (2015). Cultural stereotypes as gatekeepers: Increasing girls' interest in computer science and engineering by diversifying stereotypes. *Frontiers in Psychology, 6*, 49. https://doi.org/10.3389/fpsyg.2015.00049

Chestnut, E. K., Zhang, M. Y., & Markman, E. M. (2021). "Just as good": Learning gender stereotypes from attempts to counteract them. *Developmental Psychology, 57*(1), 114–125. https://doi.org/10.1037/dev0001143

Christensen, R., Knezek, G., & Tyler-Wood, T. (2015). Alignment of hands-on STEM engagement activities with positive STEM dispositions in secondary school students. *Journal of Science Education and Technology, 24*, 898–909. https://doi.org/10.1007/s10956-015-9572-6

Cvencek, D., Kapur, M., & Meltzoff, A. N. (2015). Math achievement, stereotypes, and math self-concepts among elementary-school students in Singapore. *Learning and Instruction, 39*, 1–10. https://doi.org/10.1016/j.learninstruc.2015.04.002

Dasgupta, N., Thiem, K. C., Coyne, A. E., Laws, H., Barbieri, M., & Wells, R. S. (2022). The impact of communal learning contexts on adolescent self-concept and achievement: Similarities and differences across race and gender. *Journal of Personality and Social Psychology, 123*(3), 537–558. https://doi.org/10.1037/pspi0000377

Degol, J. L., Wang, M. T., Zhang, Y., & Allerton, J. (2018). Do growth mindsets in math benefit females? Identifying pathways between gender, mindset, and motivation. *Journal of Youth and Adolescence, 47*, 976–990. https://doi.org/10.1007/s10964-017-0739-8

Del Toro, J., & Wang, M. T. (2023). Stereotypes in the classroom's air: Classroom racial stereotype endorsement, classroom engagement, and STEM achievement among Black and White American adolescents. *Developmental Science, 26*, e13380. https://doi.org/10.1111/desc.13380

Diekman, A. B., Brown, E. R., Johnston, A. M., & Clark, E. K. (2010). Seeking congruity between goals and roles: A new look at why women opt out of science, technology, engineering, and mathematics careers. *Psychological Science, 21*(8), 1051–1057. https://doi.org/10.1177/0956797610377342

Dweck, C. S. (2006). *Mindset: A new psychology of success*. Random House.

Falco, L. D., & Summers, J. J. (2019). Improving career decision self-efficacy and STEM self-efficacy in high school girls: Evaluation of an intervention. *Journal of Career Development, 46*(1), 62–76. https://doi.org/10.1177/0894845317721651

Gardner, K., Glassmeyer, D., & Worthy, R. (2019). Impacts of STEM professional development on teachers' knowledge, self-efficacy, and practice. *Frontiers in Education, 4*, 1–10. https://doi.org/10.3389/feduc.2019.00026

Gibson, B., & Hutton, R. (2017). Public perceptions of engineers and engineering. https://engineerscanada.ca/sites/default/files/public-perceptions-of-engineers-and-engineering.pdf

Haimovitz, K., & Dweck, C. S. (2017). The origins of children's growth and fixed mindsets: New research and a new proposal. *Child Development, 88*(6), 1849–1859. https://doi.org/10.1111/cdev.12955

Hayes, A. R., Bigler, R. S., & Weisgram, E. S. (2018). Of men and money: Characteristics of occupations that affect the gender differentiation of children's occupational interests. *Sex Roles, 78*, 775–788. https://doi.org/10.1007/s11199-017-0846-8

Herbert, J., & Stipek, D. (2005). The emergence of gender differences in children's perceptions of their academic competence. *Journal of Applied Developmental Psychology, 26*(3), 276–295. https://doi.org/10.1016/j.appdev.2005.02.007

Hughes, J. M., & Bigler, R. S. (2008). The impact of race on children's occupational aspirations. In S. M. Quintana & C. McKown (Eds.), *Handbook of race, racism, and the developing child* (pp. 397–423). John Wiley & Sons.

Katehi, L., Pearson, G., & Feder, M. (2009). *Engineering in K–12 education: Understanding the status and improving the prospects*. National Academies Press.

Kiran, D., & Sungur, S. (2012). Middle school students' science self-efficacy and its sources: Examination of gender difference. *Journal of Science Education and Technology, 21*, 619–630. https://doi.org/10.1007/s10956-011-9351-y

Kloser, M., Wilsey, M., Twohy, K. E., Immonen, A. D., & Navotas, A. C. (2018). "We do STEM": Unsettled conceptions of STEM education in middle school STEM classrooms. *School Science and Mathematics, 118*(8), 335–347. https://doi.org/10.1111/ssm.12304

Law, F., McGuire, L., Winterbottom, M., & Rutland, A. (2021). Children's gender stereotypes in STEM following a one-shot growth mindset intervention in a science museum. *Frontiers in Psychology, 12*, 641695. https://doi.org/10.3389/fpsyg.2021.641695

Leaper, C., & Brown, C. S. (2014). Sexism in schools. *Advances in Child Development and Behavior, 47*, 189–223. https://doi.org/10.1016/bs.acdb.2014.04.001

Lei, R. F., Green, E. R., Leslie, S. J., & Rhodes, M. (2019). Children lose confidence in their potential to "be scientists," but not in their capacity to "do science." *Developmental Science, 22*(6), e12837. https://doi.org/10.1111/desc.12837

Liben, L. S., & Coyle, E. F. (2014). Developmental interventions to address the STEM gender gap: Exploring intended and unintended consequences. *Advances* in *Child Development and Behavior, 47*, 77–115. https://doi.org/10.1016/bs.acdb.2014.06.001

Lie, R., Selcen Guzey, S., & Moore, T. J. (2019). Implementing engineering in diverse upper elementary and middle school science classrooms: Student learning and attitudes. *Journal of Science Education and Technology, 28*, 104–117. https://doi.org/10.1007/s10956-018-9751-3

Maltese, A. V., & Tai, R. H. (2011). Streams persistence: Examining the association of educational experiences with earned degrees in STEM among U.S. students. *Science Education, 95*(5), 877–907. https://doi.org/10.1002/sce.20441

Master, A. (2021). Gender stereotypes influence children's STEM motivation. *Child Development Perspectives, 15*(3), 203–210. https://doi.org/10.1111/cdep.12424

Master, A., Cheryan, S., & Meltzoff, A. N. (2014). Reducing adolescent girls' concerns about STEM stereotypes: When do female teachers matter? *Revue internationale de psychologie sociale, 27*(3), 79–102.

Master, A., & Meltzoff, A. N. (2016). Building bridges between psychological science and education: Cultural stereotypes, STEM, and equity. *Prospects, 46*(2), 215–234.

Master, A., & Meltzoff, A. N. (2020). Cultural stereotypes and sense of belonging contribute to gender gaps in STEM. *International Journal of Gender, Science and Technology, 12*(1), 152–198. https://doi.org/10.1111/cdep.12424

Master, A., Meltzoff, A. N., & Cheryan, S. (2021). Gender stereotypes about interests start early and cause gender disparities in computer science and engineering. *Proceedings of the National Academy of Sciences, 118*(48), e2100030118.

Meyer, M., Cimpian, A., & Leslie, S. J. (2015). Women are underrepresented in fields where success is believed to require brilliance. *Frontiers in Psychology, 6*, 235. https://doi.org/10.3389/fpsyg.2015.00235

Miller, P. H., Blessing, J., & Schwartz, S. (2006). Gender differences in high-school students' views about science. *International Journal of Science Education, 28*(4), 363–381. https://doi.org/10.1080/09500690500277664

Oppermann, E., Brunner, M., & Anders, Y. (2019). The interplay between preschool teachers' science self-efficacy beliefs, their teaching practices, and girls' and boys' early science motivation. *Learning and Individual Differences, 70*, 86–99. https://doi.org/10.1016/j.lindif.2019.01.006

Owen, W. (2011). Intel survey of teenagers shows they don't know what engineers do, limiting them from choosing those careers. *The Oregonian.* https://www.oregonlive.com/hillsboro/2011/12/intel_survey_of_teenagers_show.html

Pahlke, E., Bigler, R. S., & Green, V. A. (2010). Effects of learning about historical gender discrimination on early adolescents' occupational judgments and aspirations. *The Journal of Early Adolescence, 30*(6), 854–894.

Plante, I., De la Sablonnière, R., Aronson, J. M., & Théorêt, M. (2013). Gender stereotype endorsement and achievement-related outcomes: The role of competence beliefs and task values. *Contemporary Educational Psychology, 38*(3), 225–235. https://doi.org/10.1016/j.cedpsych.2013.03.004

Porter, T., Catalán Molina, D., Cimpian, A., Roberts, S., Fredericks, A., Blackwell, L. S., & Trzesniewski, K. (2022). Growth-mindset intervention delivered by teachers boosts achievement in early adolescence. *Psychological Science, 33*(7), 1086–1096. https://doi.org/10.1177/09567976211061109

Rhodes, M., Leslie, S. J., Yee, K. M., & Saunders, K. (2019). Subtle linguistic cues increase girls' engagement in science. *Psychological Science, 30*(3), 455–466. https://doi.org/10.1177/0956797618823670

Rubie-Davies, C. M., & Peterson, E. R. (2010). Teacher expectations and beliefs: Influences on the socioemotional environment of the classroom. In C. M. Rubie-Davies (Ed.), *Educational psychology: Concepts, research and challenges* (pp. 148–163). Routledge.

Shin, J. E. L., Levy, S. R., & London, B. (2016). Effects of role model exposure on STEM and non-STEM student engagement. *Journal of Applied Social Psychology, 46*(7), 410–427. https://doi.org/10.1111/jasp.12371

Smith, J. L., Cech, E., Metz, A., Huntoon, M., & Moyer, C. (2014). Giving back or giving up: Native American student experiences in science and engineering. *Cultural Diversity and Ethnic Minority Psychology, 20*(3), 413–429. https://doi.org/10.1037/a0036945

Starr, C. R., & Simpkins, S. D. (2021). High school students' math and science gender stereotypes: Relations with their STEM outcomes and socializers' stereotypes. *Social Psychology of Education, 24*, 273–298. https://doi.org/10.1007/s11218-021-09611-4

Stohlmann, M. (2022). Growth mindset in K-8 STEM education: A review of the literature since 2007. *Journal of Pedagogical Research, 6*(2), 149–163. https://dx.doi.org/10.33902/JPR.202213029

Stout, J. G., Dasgupta, N., Hunsinger, M., & McManus, M. A. (2011). STEMing the tide: Using ingroup experts to inoculate women's self-concept in Science, Technology, Engineering, and Mathematics (STEM). *Journal of Personality and Social Psychology, 100*(2), 255–270. https://doi.org/10.1037/a0021385

Thoman, D. B., Brown, E. R., Mason, A. Z., Harmsen, A. G., & Smith, J. L. (2015). The role of altruistic values in motivating underrepresented minority students for biomedicine. *BioScience, 65*(2), 183–188. https://doi.org/10.1093/biosci/biu199

Thomas, D. R., & Larwin, K. H. (2023). A meta-analytic investigation of the impact of middle school STEM education: Where are all the students of color?. *International Journal of STEM Education, 10*, 43. https://doi.org/10.1186/s40594-023-00425-8

Wang, M. M., Cardarelli, A., Leslie, S. J., & Rhodes, M. (2022). How children's media and teachers communicate exclusive and essentialist views of science and scientists. *Developmental Psychology, 58*(8), 1455–1471. https://doi.org/10.1037/dev0001364

Wang, M. T., & Degol, J. L. (2017). Gender gap in Science, Technology, Engineering, and Mathematics (STEM): Current knowledge, implications for practice, policy, and future directions. *Educational Psychology Review, 29*, 119–140. https://doi.org/10.1007/s10648-015-9355-x

Way, N., Hernández, M. G., Rogers, L. O., & Hughes, D. L. (2013). "I'm not going to become no rapper": Stereotypes as a context of ethnic and racial identity development. *Journal of Adolescent Research, 28*(4), 407–430. https://doi.org/10.1177/0743558413480836

Wiebe, E. N., Faber, M., & Corn, J., Collins, T. L., Unfried, A., & Townsend, L. (2013, June). *A large-scale survey of K–12 students about STEM: Implications for engineering curriculum development and outreach efforts (Research to practice)*. Paper presented at 2013 ASEE Annual Conference & Exposition, Atlanta, Georgia. https://doi.org/10.18260/1-2--19073

Weisgram, E. S., & Bigler, R. S. (2006). Girls and science careers: The role of altruistic values and attitudes about scientific tasks. *Journal of Applied Developmental Psychology, 27*(4), 326–348. https://doi.org/10.1016/j.appdev.2006.04.004

Weisgram, E. S., & Bigler, R. S. (2007). Effects of learning about gender discrimination on adolescent girls' attitudes toward and interest in science. *Psychology of Women Quarterly, 31*, 262269. https://doi.org/10.1111/j.1471-6402.2007.00369.x

Weisgram, E. S., Dinella, L. M., & Fulcher, M., (2011). The role of masculinity/femininity, values, and occupational value affordances in shaping young men's and women's occupational choices. *Sex Roles, 65*(3), 243–258. https://doi.org/10.1007/s11199-011-9998-0

Wilson, D., Bates, R., Scott, E. P., Painter, S. M., & Shaffer, J. (2015). Differences in self-efficacy among women and minorities in STEM. *Journal of Women and Minorities in Science and Engineering, 21*(1), 27–45. 10.1615/JWomenMinorScienEng.2014005111

Yeager, D. S., & Dweck, C. S. (2020). What can be learned from growth mindset controversies? *American Psychologist, 75*(9), 1269–1284. https://doi.org/10.1037/amp0000794

Chapter 8

Beyond the Technocratic View of Engineering

Gillian Roehrig

The release of STEM policy documents throughout the world has generated renewed focus on interdisciplinary instruction. Indeed, recent systemic reviews show exponential growth in the number of STEM (and STEAM) education publications over the past decade (Zhao, 2022; Li et al., 2020). While integrated STEM education is well established, disagreement on models and approaches persists (Moore et al., 2020). Indeed, Sgro, Bobowski, and Oliveira (2020, p. 185) argue that, in essence, integrated STEM is "whatever someone decides it means," creating a need for "greater clarity about not only what constitutes STEM education, but how educators as a whole conceptualize STEM and the process of integration." This work is further complicated by the more recent development of approaches such as STEAM, STREAM, and STREAMS education, an issue that is interrogated throughout the chapters in this book.

While some researchers argue for integration across all four of the STEM disciplines (e.g., Burrows et al., 2018), most definitions of integrated STEM only call for the inclusion of at least two of the STEM disciplines (e.g., Roehrig et al., 2021; Kelley & Knowles, 2016). Within the United States, the *Framework for K–12 Science Education* (National Research Council (NRC) 2012) and the *Next Generation Science Standards* (NGSS Lead States, 2013) place the responsibility for integration of engineering and engineering practices on science teachers. Consequently, the preponderance of integrated STEM research occurs within the context of science education (Takeuchi et al., 2020) and most often uses an engineering context or engineering design problem as central to integrated STEM education (e.g., Roehrig et al., 2021; Berland & Steingut, 2016; Moore et al., 2014a). Indeed, the most common combination of STEM disciplines within integrated STEM education is science and engineering (Moore et al., 2020). For example, Calabrese Barton and Tan (2019, p. 2) specifically defined STEM as "the integration of engineering into science learning goals and experiences as outlined by the Next Generation Science Standards."

Given the increasingly prominent role of engineering in K–12 science education, this chapter explores the role of different disciplines in the effective implementation of K–12 engineering. First, integrated STEM and the central role of engineering are described. Second, the technical emphasis of integrated STEM is problematized, leading to an exploration of STEAM, STEAMS, and STREAMS education.

INTEGRATED STEM

The primary policy driver for integrated STEM education is the argument that national prosperity depends on addressing growing shortages in the STEM workforce (e.g., President's Council of Advisors on Science and Technology, 2011; National Academy of Science, National Academy of Engineering, and Institute of Medicine, 2007). However, educators and researchers have attempted to "refocus the aims of STEM education from pursuing economic profits to responding to the fundamental challenges that humanity is faced with" (Park et al., 2022, p. 4). Researchers argue that integrated STEM should foster social and environmental justice (Mejias et al., 2019) as part of the goal to develop STEM literacy and awareness (Bybee, 2013).

While some debate continues about a definition for integrated STEM, there is strong consensus in the literature on characteristics of integrated STEM: (a) the use of real-world problems to contextualize students learning (e.g., Kloser et al., 2018; Kelley & Knowles, 2016), (b) connections between STEM disciplines should be made explicit to students (e.g., English, 2016; Kelley & Knowles, 2016), (c) the development of twenty-first-century skills (e.g., Wang & Knoblach, 2018; Sias et al., 2017), and (d) the use of student-centered pedagogies (e.g., Thibaut et al., 2018; Johnson et al., 2016).

Focus on Real-world Problems

The most common characteristic of integrated STEM in the literature is that STEM integration should be contextualized by a real-world problem (e.g., Moore et al., 2020; Kloser et al., 2018; Kelley & Knowles, 2016). Real-world contexts provide motivation and purpose for learning STEM content and can improve student interest in science and engineering (McLure et al., 2021; Lachapelle & Cunningham, 2014). Given that real-world problems are inherently multidisciplinary in nature, developing solutions to real-world problems relies on using and developing an understanding of content from multiple disciplines (e.g., Thibaut et al., 2018; Cavlazoglu & Stuessy, 2017). However, there are important considerations about the nature of such problems if student learning and motivation are to be promoted for all students (Roehrig et al., 2021).

While several approaches to addressing real-world problems exist, many researchers and educators use engineering design tasks as a vehicle for bounding the scope and focus of real-world problems (Roehrig et al., 2021; Moore et al., 2020). Engineering addresses the needs of a client through a systematic and iterative approach to designing solutions to real-world problems (NRC 2012). Engineering design is the central activity of engineering (Dym, 1999), and as such current reforms in the United States call for students to engage in the engineering design process (NGSS Lead States, 2013; NRC, 2012). Within policy documents, engineering design is an iterative process of "testing the most promising solutions and modifying what is proposed on the basis of the test results leads to greater refinement and ultimately to an optimal solution" (NRC, 2012, p. 210).

Connections across the Disciplines

Engineers apply science, mathematics, and engineering knowledge when designing solutions, and K–12 students should "have the opportunity to apply developmentally appropriate mathematics or science in the context of solving engineering problems" (Moore et al., 2014b, p. 5). Unfortunately, many implementations of K–12 engineering activities negate the application of STEM content, becoming little more than craft projects, with students resorting to tinkering, without thoughtful consideration of STEM content (McComas & Burgin, 2020). Thus, successful implementation of integrated STEM necessitates an authentic problem or engineering design challenge that requires students to *explicitly* learn and apply science and mathematics concepts.

STEM Practices and Twenty-First-Century Skills

Common across definitions of integrated STEM are expectations that students engage in STEM practices and twenty-first-century skills (e.g., critical thinking, creativity, communication, and collaboration) (e.g., Sias et al., 2017; Kelley & Knowles, 2016; Moore et al., 2014a). Real-world problems and engineering design challenges are complex with multiple solution pathways, which promotes students' creativity and critical thinking in determining the viability of different possible solutions (Stretch & Roehrig, 2021; Simpson et al., 2018). As engineers iteratively test and improve their design solutions, it is inevitable that early prototypes will not fully address the criteria and constraints imposed by the client. Indeed, failure is expected, and it is through creativity and critical thinking that these failures lead to stronger designs and innovation (Henry et al., 2021; Simpson et al., 2018). Indeed, engineering requires students to be "independent, reflective, and metacognitive thinkers who understand that prior experience and learning from failure can ultimately lead to better solutions" (Moore et al., 2014b, p. 5).

Student Agency

The use of student-centered pedagogies allows students to have agency in design decisions as they engage in the engineering design process (e.g., Berland & Steingut, 2016; Johnson et al., 2016). The multiple solution pathways possible within a real-world problem or engineering design task allow students to determine their own solution trajectories and opportunities to build knowledge. Thus, students are positioned as epistemic agents as opposed to receivers of STEM content, allowing them to propose solutions to personally meaningful problems rather than simply learning the canonical facts of the discipline (Miller et al., 2018). Miller et al. (2018) define epistemic agency as "students being positioned with, perceiving, and acting on, opportunities to shape the knowledge building work in their classroom community" (p. 1058).

The notion of epistemic agency is particularly relevant when considering students who are traditionally underrepresented in STEM. In order to engage underrepresented students in STEM, educators need to find ways to promote positive STEM identities (Thomas & Williams, 2010). Integrated approaches, such as STEM and STEAM, have been identified as a way to promote the development of a STEM-related identity and thus promote a more diverse STEM workforce (National Academy of Engineering and National Research Council, 2014). Epistemic agency promotes students using and building on personal and cultural knowledge, as well as canonical STEM knowledge, to propose solutions to personally meaningful problems (Schwarz et al., 2017). By allowing students to have epistemic agency related to personally relevant issues that directly help others, underrepresented students are more likely to develop a positive STEM identity and consider STEM as a future career path (Billington et al., 2013; Diekman et al., 2015; Harackiewicz et al., 2016; Miller et al., 2018; Weisgram et al., 2010).

Critiques of Integrated STEM

Unfortunately, K–12 students usually experience engineering design solely as a technical problem (Gunckel & Tolbert, 2018), with criteria and constraints limited to issues such as time, access to materials, and budget. While such constraints are realistic, this technocentric approach ignores the social, political, and ethical issues that are inherent in most real-world problems (Roehrig et al., 2020; Gunckel & Tolbert, 2018). However, engineers must be aware of the societal and environmental impacts of their work (DeVere et al., 2009). Thus, engineering scholars recommend that the engineering profession must evolve from solely technical to a profession that is guided by its "understanding of the human, environmental, societal and cultural challenges and the consequences of professional activity" (DeVere et al., 2009, p. 1). Thus, the Accreditation Board for Engineering and Technology (ABET) calls for the inclusion of social science and humanities into the undergraduate engineering curriculum, and similarly, engineering educators call for the inclusion of consideration of societal impacts (Moore et al., 2014b). The development of STEM-literate citizens and a future STEM workforce who think beyond the traditional technocratic focus necessitates attention to the role of the social sciences in engineering (Roehrig et al., 2020; Gunckel & Tolbert, 2018). These concerns about the technocentric focus of integrated STEM have led to calls for new interdisciplinary approaches integrating the arts and social sciences.

FROM STEM TO S(R)TEAM(S)

Science, Technology, Engineering, Arts, and Mathematics (STEAM) education was first conceptualized in 2006 by Georgette Yakman, and while STEAM articles only represent 10 percent of the STEM education publications from 2016 to 2021 (Zhao, 2023), there has been growing international interest in STEAM education (e.g., Belbase et al., 2021; Quigley & Herro, 2016). Rather than positioning STEAM as a new construct, many researchers treat it "as an extension of [integrated] STEM for driving economic and national competitiveness" (Mejias et al., 2019, p. 210). Thus, two dominant approaches to STEAM education include perceiving the integration of the arts as (i) promoting learning in STEM disciplines (Ge et al., 2015) and (ii) enhancing students' general skills, such as perspective-taking, creative and problem-solving skills (Perignat & Katz-Buonincontro, 2019). Like efforts in integrated STEM education to move beyond the singular goal of workforce development, STEAM education is also viewed as offering "new ways of doing and knowing in the arts and STEM fields, often with emancipatory and critical pedagogical approaches to learning that can be at odds with an economistic focus" (Mejias et al., 2019, p. 210). A third approach to STEAM education represents a political move to re-emphasize the arts, using STEAM as a vehicle for supporting arts education (Ge et al., 2015).

Despite its increasing popularity, STEAM education suffers from a lack of conceptual clarity (Aguilera & Ortiz Revilla, 2021; Matsuura & Nakamura,

2021) as evidenced by the diverse purposes described above. This lack of clarity is also evident in disagreements about the definition of STEAM itself, particularly with regard to the nature of the A in STEAM, including A as the visual arts (e.g., drawing, sculpture, and photography), A as the performing arts (e.g., dance, theatre, and music), A as the liberal arts and humanities (Quigley et al., 2017; National Art Education Association, 2016), and in some cases A as agriculture (e.g., Reyes et al., 2021; Sumida, 2017; Toni, 2014). While agriculture can provide a relevant context for engineering design tasks, this chapter focuses on the arts as providing potential benefits to students' engagement in engineering and STEM activities. Some approaches are focused on enhancing the learning of STEM concepts, for example, improving memorization and visualization of complex ideas (e.g., Thomson & Sefton-Green, 2011); however, other approaches are more specific to engineering. The inclusion of the arts as relevant for STEM and engineering education is portrayed in the literature as adding a focus on aesthetics to engineering design (Bequette & Bequette, 2015), injecting creativity and innovation absent from STEM (Segarra et al., 2018), and maker spaces (Taylor, 2016).

Engineers and artists engage in design thinking in their work. Indeed, the engineering design process shares many features with functional design as taught in the arts (Bequette & Bequette, 2015). Vande Zande (2010) argues that commercial projects require a balance of function and aesthetics. Bequette and Bequette (2015, p. 44) argue that the "design of products, buildings, computer graphics, interactive video games, and the like is thus more aesthetically grounded and artistically motivated than is apparent in an engineering design cycle." Depending on the nature of the engineering design challenge (e.g., developing a process for sorting recycled trash vs. developing a video game), the infusion of the arts could add value to the engineering design process with respect to aesthetics.

Common across definitions of STEAM is the role of creativity. For example, Wannapiroon and Petsangsri (2020, p. 1648) define STEAM Education as integrating "science, technology, engineering, art, and mathematics in order to provide learners with creative skill, investigation skill, debate skill, critical thinking skill, and creativity and innovation." Similarly, Conradty and Bogner (2020, p. 1) define STEAM as "the integration of arts (A) and creativity in the classical STEM teaching." Such definitions are problematic as they position creativity as existing solely within the domain of the arts. Creativity is essential to the field of engineering; indeed, the success of an engineer depends on creativity and innovation (Zhou, 2012). A creative engineer should be able to "explore and scrutinize the available data or information and generate novel solutions to specific engineering problems or to the production of a unique product" (Liu & Schonwetter, 2004, p. 801). Within engineering, a product needs to be both novel and functional, with a focus on functional creativity (Charyton & Merrill, 2009) given that the product needs to effectively perform the defined task, as well as being able to be economically and safely manufactured. Whereas, in artistic creativity novelty alone can suffice, as "effectiveness may be judged according to purely aesthetic or expressive criteria" (Hoffmann et al., 2005, p. 163). There exists a body of research on creativity in engineering that explores the development and implementation of creativity in both engineering and engineering education (e.g., Zhou, 2012; Liu & Schonwetter, 2004) and as such, given the different goals of engineers and artists, the role of the arts in enhancing creativity in engineering would benefit from more clarity.

Makerspaces have proliferated in K–12 schools over the past ten years (Mejias et al., 2019; Taylor, 2016). Most makerspaces resemble an arts studio or a modernized school shop class, including access to wood and metal-working tools, 3-D printers, laser cutters, sewing machines, and kilns (McComas & Burgin, 2020; Blackley et al., 2017) with the goal of "creative production in art, science, and engineering" that "blend digital and physical technologies to explore ideas, learn technical skills, and create new products" (Sheridan et al., 2014, p. 505). Unfortunately, the focus on product development usually lacks any connection to a real-world problem or societal need (McComas & Burgin, 2020), leading to engagement in tinkering with limited learning and application of science or mathematics content within makerspace settings (Sheffield et al., 2017). In other words, makerspaces engage students in engineering-like activities rather than authentic engineering practices aligned with characteristics of quality K–12 engineering and integrated STEM education (e.g., Roehrig et al., 2021; Sheffield et al., 2017; Moore et al., 2014b).

STEAM(S) EDUCATION

Most commensurate with complementing K–12 integrated STEM activities using a central engineering design activity is the representation of the A in STEAM as the liberal arts and humanities (Quigley et al., 2017; National Art Education Association, 2016), sometimes also represented as STEAMS with the inclusion of an additional S (social sciences). Attention to the social sciences directly addresses the critique that integrated STEM is primarily implemented with a technocentric approach that ignores critical social, political, and ethical issues (Roehrig et al., 2020; Gunckel & Tolbert, 2018). Indeed, "In an increasingly complex world, there exists a need for an interdisciplinary, [humanistic] approach to engineering that includes the expansion of a dialogue that integrates the humanities and the social sciences with engineering" (Grasso et al., 2004, p. 413). Engineers must be cognizant of the cultural and environmental impacts of their work and its impact on society and quality of life (DeVere et al., 2009).

This tendency in K–12 settings to focus on the technical aspects of engineering also "leads to a utilitarian belief that technological process and solutions are inherently good and progressive, without considering who is deciding on what is good or beneficial, often privileging the majority culture or group" (McCurdy et al., 2020, p. 27). This has led to a call for a focus on empathy in STEM and engineering (e.g., McCurdy et al., 2020; Gunckel & Tolbert, 2018). McCurdy et al. (2020, p. 26) define empathy as "the stimulus that connects students to the person or cause for which their solution aims to benefit." Not only does attention to empathy create a focus on societal and cultural impacts, in addition to technical aspects of a real-world problem, it also addresses equity issues in the field of engineering. Attention to empathy, or an ethic of care, is a focus of research attending to improving gender and racial equity in engineering and the STEM fields (Leammukda et al., 2023; McCurdy et al., 2020; Burns & Lesseig, 2017; Capobianco & Yu, 2014).

While there is a clear need for attention to the social sciences (whether represented in the acronym as A or an additional S), it is also evident that engineering scholars view this as part of engineering thinking (e.g., DeVere et al., 2009; Grasso et al., 2004). More complete understandings of engineering and authentic engagements in engineering practices would address the technocentric critique of K–12 STEM and engineering activities without the inherent lack of clarity of moving from STEM to STEAM(S). The focus within the STEM and engineering education fields is on designing curricular tasks that promote engagements in aspects of engineering related to empathy and attention to societal and cultural influences and impacts (e.g., Leammukda et al., 2023; McCurdy et al., 2020; Burns & Lesseig, 2017; Capobianco & Yu, 2014) rather than STEAM(S) approaches with their built-in lack of conceptual clarity.

STREAM(S) EDUCATION

Evolving beyond STEAM education is the nascent field of STREAM(s) education, which primarily includes R as reading and writing (e.g., Clements et al., 2020; Nuangchalerm et al., 2020). However, R is also represented as recreation (Yoh et al., 2021), robotics (Badmus & Omosewo, 2020), reality (Opriş et al., 2021), and reflexive deliberation (Krug and Shaw 2016), adding further lack of clarity compared to STEAM(S) education. Unquestionably, engineering, like any other discipline, requires attention to reading and writing. For example, in their Framework for Quality K–12 Engineering Education, Moore et al. (2014b) include engineering communication as a key indicator. They describe the need to address technical writing to explain both design and process and communication in everyday language for those without an engineering background. Students would be expected to engage in communication activities such as writing client reports and creating presentations. Additionally, the K–12 engineering curriculum sometimes uses storybooks as a vehicle to introduce a client and their needs, particularly in the elementary grades (e.g., Cunningham, 2009).

CONCLUSIONS

While there is a clear need to move beyond technocentric integrated STEM and engineering lessons in K–12 classrooms, in order for STEAM, STEAMS, or STREAMS approaches to have a meaningful impact on students will take more clarity and clearer conceptualization. Indeed, many arguments for STEAM education mirror the workforce arguments that were the driver for the integration of engineering into K–12 science classrooms. So, it is important to first have a clear understanding of the role of "the arts" in enhancing engineering and/or STEM education. Defining the A as the liberal arts or humanities has the strongest potential to enhance the limited enactments of engineering as defined in current policy documents (e.g., NGSS Lead States, 2013; NRC, 2012). However, a more complete representation of engineering and engineering practices would include the "ability to apply engineering design to produce solutions that meet specified needs with consideration of public health, safety, and welfare, as well as global, cultural, social, environmental, and economic factors" (ABET, 2022). Engineering is no longer a purely technical discipline; rather, approaches such as "sustainable design, human-centered design, value-sensitive design, and universal design have gained popularity in pushing engineering design toward a more holistic thought process" (Washuta et al., 2022, p. 2). These approaches call for engineers to consider effects on the environment and natural resources and to design for the public good. Granted these approaches draw on the humanities but representing engineering design in K–12 classrooms through these approaches instead of a traditional technical approach would address critiques of current implementations of K–12 engineering. The addition of "the arts" broadly as the humanities into STEM could support a societally aware approach to engineering; however, without significant conceptualization, a performing arts addition into STEM does not yet support the goal of best representing real-world engineering in K–12 schools.

REFERENCES

ABET. (2022). Criteria for accrediting engineering programs. https://www.abet.org/accreditation/accreditation-criteria/criteria-for-accrediting- engineering-programs-2022-2023/

Aguilera, D., & Ortiz-Revilla, J. (2021). STEM vs. STEAM education and student creativity: A systematic literature review. *Education Sciences, 11*(7), 331.

Badmus, O., & Omosewo, E. O. (2020). Evolution of STEM, STEAM and STREAM education in Africa: The implication of the knowledge gap. *International Journal on Research in STEM Education, 2*(2), 99–106.

Berland, L. K., & Steingut, R. (2016). Explaining variation in student efforts toward using math and science knowledge in engineering contexts. *International Journal of Science Education, 38*(18), 2742–2761.

Blackley, S., Sheffield, R., Maynard, N., Koul, R., & Walker, R. (2017). Makerspace and reflective practice: Advancing pre-service teachers in STEM education. *Australian Journal of Teacher Education, 42*(3), 22–37.

Burrows, A., Lockwood, M., Borowczak, M., Janak, E., & Barber, B. (2018). Integrated STEM: Focus on informal education and community collaboration through engineering. *Education Sciences, 8*(4), 1–15. http://doi.org/10.3390/educsci8010004

Charyton, C., & Merrill, J. A. (2009). Assessing general creativity and creative engineering design in first year engineering students. *Journal of Engineering Education, 98*(2), 145–156.

Conradty, C., & Bogner, F. X. (2020). STEAM teaching professional development works: Effects on students' creativity and motivation. *Smart Learning Environments, 7*, 1–20.

Cunningham, C. M. (2009). Engineering is elementary. *The Bridge, 30*(3), 11–17.

Belbase, S., Mainali, B. R., Kasemsukpipat, W., Tairab, H., Gochoo, M., & Jarrah, A. (2021). At the dawn of science, technology, engineering, arts, and mathematics (STEAM) education: Prospects, priorities, processes, and problems. *International Journal of Mathematical Education in Science and Technology, 53*, 2919–2955.

Bequette, J. W., & Bequette, M. B. (2015). A place for art and design education in the STEM conversation. *Art Education, 65*(2), 40–47.

Burns, H. D., & Lesseig, K. (2017, June). Infusing empathy into engineering design: Supporting under-represented student interest and sense of belongingness. In *2017 ASEE Annual Conference & Exposition*.

Bybee, R. W. (2013). *A case for STEM education*. National Science Teachers' Association Press.

Calabrese Barton, A., & Tan, E. (2019). Designing for rightful presence in STEM: The role of making present practices. *Journal of the Learning Sciences, 28*(4–5), 616–658.

Capobianco B., & Yu, J. H. (2014). Using the construct of care to frame engineering as a caring profession toward promoting young girls' participation. *Journal of Women and Minorities in Science and Engineering, 20*, 21–33.

Cavlazoglu, B., & Stuessy, C. L. (2017). Identifying and verifying earthquake engineering concepts to create a knowledge base in STEM education: A modified Delphi study. *International Journal of Education in Mathematics, Science and Technology, 5*(1), 40–52.

Clements, D. H., Sarama, J., Brenneman, K., Duke, N. K., & Hemmeter, M. L. (2020). STREAM ēducation at work—no, at play!. *YC Young Children, 75*(2), 36–43.

DeVere, I., Johnson, K. B., & Thong, C. (2009). *Educating the responsible engineer: Socially responsible design and sustainability in the curriculum.* International conference on engineering and product design education 10 & 11 September 2009, University of Brighton, UK.

Dym, C. (1999). Learning engineering: Design, languages, and experiences. *Journal of Engineering Education, 88*(2), 145–148.

English, L. D. (2016). STEM education K-12: Perspectives on integration. *International Journal of STEM Education, 3*(1), 1–8.

Ge, X., Ifenthaler, D., & Spector, J. (2015). Moving forward with STEAM education research. In X. Ge, D. Ifenthaler, & J. Spector (Eds.), *Emerging technologies for STEAM education. Educational communications and technology: Issues and innovations* (pp. 383–396). Springer.

Grasso, D., Callahan, M., & Doucett, S. (2004). Defining engineering thought. *International Journal of Engineering Education, 20*(3), 412–415.

Gunckel, K. L., & Tolbert, S. (2018). The imperative to move toward a dimension of care in engineering education. *Journal of Research in Science Teaching, 55*(7), 938–961.

Henry, M. A., Shorter, S., Charkoudian, L. K., Heemstra, J. M., Le, B., & Corwin, L. A. (2021). Quantifying fear of failure in STEM: Modifying and evaluating the Performance Failure Appraisal Inventory (PFAI) for use with STEM undergraduates. *International Journal of STEM Education, 8*(43). https://doi.org/10.1186/s40594-021-00300-4

Hoffmann, O., Cropley, D., Cropley, A., Nguyen, L. & Swatman, P. (2005). Creativity, requirements and perspectives, *Australasian Journal of Information Systems, 13*(1), 159–175.

Johnson, C. C., Peters-Burton, E. E., & Moore, T. J. (2016). *STEM road map: A framework for integrated STEM education.* Routledge.

Kelley, T. R., & Knowles, J. G. (2016). A conceptual framework for integrated STEM education. *International Journal of STEM Education, 3*(1), 1–11.

Kloser, M., Wilsey, M., Twohy, K. E., Immonen, A. D., & Navotas, A. C. (2018). "We do STEM": Unsettled conceptions of STEM education in middle school S.T.E.M. classrooms. *School Science & Mathematics, 118*(8), 335–347.

Krug, D., & Shaw, A. (2016). Reconceptualizing ST®E(A)M(S) education for teacher education. *Canadian Journal of Science, Mathematics and Technology Education, 16*, 183–200.

Lachapelle, C., & Cunningham, C. (2014). Engineering in elementary schools. In S. Purzer, J. Strobel, & M. Cardella (Eds.), *Engineering in pre-college settings: Synthesizing research, policy, and practices* (pp. 61–88). Purdue University Press.

Leammukda, F., Boyd, B., & Roehrig, G. (2023). Fostering STEM interest in middle school girls through community-embedded integrated STEM. *Journal of Women and Minorities in Science and Engineering, 30*(2), 59–87. 10.1615/ JWomenMinorScienEng.2023039905

Li, Y., Wang, K., Xiao, Y., & Froyd, J. (2020). Research and trends in STEM education: A systematic review of journal publications. *International Journal of STEM Education, 7*(1), 1–16. https://doi.org/10.1186/s40594-020-00207-6

Liu, Z., & Schonwetter, D. J. (2004). Teaching creativity in engineering. *International Journal of Engineering Education, 20*(5), 801–808.

Matsuura, T., & Nakamura, D. (2021). Trends in STEM/STEAM education and students' perceptions in Japan. *Asia-Pacific Science Education, 7*(1), 7–33.

McLure, F. I., Koul, R. B., & Fraser, B. J. (2021). Gender differences among students undertaking iSTEM projects in multidisciplinary vs. uni-disciplinary STEM classrooms in government vs. non-government schools: Classroom emotional climate and attitudes. *Learning Environments Research.* https://doi.org/10.1007/s10984-021-09392-9

McComas, W. F., & Burgin, S. R. (2020). A critique of "STEM" education revolution-in-the-making, passing fad, or instructional imperative? *Science & Education, 29,* 805–829.

McCurdy, R. P., Nickels, M. L., & Bush, S. B. (2020). Problem-based design thinking tasks: Engaging student empathy in STEM. *The Electronic Journal for Research in Science & Mathematics Education, 24*(2), 22–55.

Mejias, S., Thompson, N., Sedas, R. M., Rosin, M., Soep, E., Peppler, K., & Bevan, B. (2021). The trouble with STEAM and why we use it anyway. *Science Education, 105*(2), 209–231.

Miller, E., Manz, E., Russ, R., Stroupe, D., & Berland, L. (2018). Addressing the epistemic elephant in the room: Epistemic agency and the next generation science standards. *Journal of Research in Science Teaching, 55*(7), 1053–1075.

Moore, T. J., Johnston, A. C., & Glancy, A. W. (2020). STEM integration: A synthesis of conceptual frameworks and definitions. In C.C. Johnson, M. J. Mohr-Schroeder, T. J. Moore, & L. D. English (Eds.), *Handbook of research on STEM education* (pp. 3–16). Routledge.

Moore, T. J., Stohlmann, M. S., Wang, H.-H., Tank, K. M., Glancy, A., & Roehrig, G. H. (2014). Implementation and integration of engineering in K-12 STEM education. In J. Strobel, S. Purzer, & M. Cardella (Eds.), *Engineering in precollege settings: Research into practice*. Sense Publishers.

Moore, T. J., Glancy, A. W., Tank, K. M., Kersten, J. A., & Smith, K. A. (2014). A framework for quality K-12 engineering education: Research and development. *Journal of Pre-College Engineering Education Research, 4*(1), 1–13.

National Academy of Sciences, National Academy of Engineering, and Institute of Medicine of the National Academies. (2007). *Rising above the gathering storm: Energizing and employing America for a brighter economic future*. National Academies Press.

National Art Education Association (NAEA). (2016). *Using art education to build a stronger workforce.* https://arteducators-prod.s3.amazonaws.com/documents/535/ff8bfae5-6b4f-4352-b900-4fc1182ad2b1.pdf?1455134278

National Research Council. (2012). *A framework for K-12 science education: Practices, crosscutting concepts, and core ideas.* National Academies Press.

NGSS Lead States. (2013). *Next generation science standards: For states, by states.* National Academies Press.

Nuangchalerm, P., Prachagool, V., Prommaboon, T., Juhji, J., Imroatun, I., & Khaeroni, K. (2020). Views of primary Thai teachers toward STREAM education. *International Journal of Evaluation and Research in Education, 9*(4), 987–992.

Opriş, I., Gogoaşe Nistoran, D. E., Costinaş, S., & Ionescu, C. S. (2021). Rethinking power engineering education for Generation Z. *Computer Applications in Engineering Education, 29*(1), 287–305.

Park, W., Wu, J. Y., & Erduran, S. (2020). The nature of STEM disciplines in the science education standards documents from the USA, Korea and Taiwan: Focusing on disciplinary aims, values and practices. *Science & Education, 29*, 899–927.

Perignat, E., & Katz-Buonincontro, J. (2019). STEAM in practice and research: An integrative literature review. *Thinking Skills and Creativity, 31*, 31–43.

President's Council of Advisors on Science and Technology. (2011). *Report to the president: Prepare and inspire: K-12 education in Science, Technology, Engineering, and Mathematics (STEM) for America's future.* Executive Office of the President.

Quigley, C. F., & Herro, D. (2016). "Finding the joy in the unknown": Implementation of STEAM teaching practices in middle school science and math classrooms. *Journal of Science Education and Technology, 25*(3), 410–426.

Quigley, C. F., Herro, D., & Jamil, F. M. (2017). Developing a conceptual model of STEAM teaching practices. *School Science and Mathematics, 117*(1–2), 1–12.

Reyes, W., Bayten, E., & Mercado, F. (2021). Differences in gender perceptions of Higher Education Institutions (HEI) students and teachers on Science Technology Engineering Agriculture Mathematics (STEAM) education. *The Normal Lights, 15*(2). https://doi.org/10.56278/tnl.v15i2.1750

Roehrig, G. H., Dare, E. A., Ellis, J. A., & Ring-Whalen, E. (2021). Beyond the basics: A detailed conceptual framework of integrated STEM. *Disciplinary and Interdisciplinary Science Education Research, 3*(1), 1–18.

Roehrig, G. H., Keratithamkul, K., & Hiwatig, B. (2020). Intersections of integrated STEM and socio-scientific issues. In W. Powell (Ed.), *Socioscientific issues-based instruction for scientific literacy development.* IGI Global.

Segarra, V. A., Natalizio, B., Falkenberg, C. V., Pulford, S., Holmes, R. M. (2018). STEAM: Using the arts to train well-rounded and creative scientists. *Journal of Microbiology & Biology education, 19*(1), 1–7.

Sgro, C. M., Bobowski, T., & Oliveira, A. W. (2020). Current praxis and conceptualization of STEM education: A call for greater clarity in integrated curriculum development. In V. Akerson & G. Buck (Eds.), *Contemporary trends and issues in science education: Critical questions in STEM education* (pp. 185–210). Springer.

Sheffield, R., Koul, R., Blackley, S., & Maynard, N. (2017). Makerspace in STEM for girls: A physical space to develop 21st century skills. *Educational Media International, 54*(2), 148–164.

Sheridan, K. M., Halverson, E. R., Litts, B. K., Brahms, L., Jacobs-Priebe, L., & Owens, T. (2014). Learning in the making: A comparative case study of three makerspaces. *Harvard Educational Review, 84*(4), 505–532.

Sias, C. M., Nadelson, L. S., Juth, S. M., & Seifert, A. L. (2017). The best laid plans: Educational innovation in elementary teacher generated integrated STEM lesson plans. *The Journal of Educational Research, 110*(3), 227–238.

Simpson, A., Anderson, A., & Maltese, A. (2019). Caught on Camera: Youth and educators' noticing of and responding to failure within making contexts. *Journal of Science Education and Technology, 28*(5), 480–492.

Stretch, E. J., & Roehrig, G. H. (2021). Framing failure: Leveraging uncertainty to launch creativity in STEM education. *International Journal of Learning and Teaching, 7*(2), 123–133.

Sumida, M. (2017). STEAM (Science, Technology, Engineering, Agriculture, and Mathematics) education for gifted young children: A glocal approach to science education for gifted young children. In *Teaching gifted learners in STEM subjects* (pp. 223–241). Routledge.

Takeuchi, M. A., Sengupta, P., Shanahan, M.-C., Adams, J. D., & Hachem, M. (2020). Transdisciplinarity in STEM education: A critical review. *Studies in Science Education, 56*(2), 213–253.

Taylor, B. (2016). Evaluating the benefit of the maker movement in K-12 STEM education. *Electronic International Journal of Education, Arts, and Science, 2*, 1–22.

Thibaut, L., Knipprath, H., Dehaene, W., & Depaepe, F. (2018a). How school context and personal factors relate to teachers' attitudes toward teaching integrated STEM. *International Journal of Technology & Design Education, 28*(3), 631–651.

Thomson, P., & Sefton-Green, J. (2011). *Researching creative learning*. Routledge.

Toni, B. (Ed.). (2014). *New Frontiers of multidisciplinary research in STEAM-H (Science, Technology, Engineering, Agriculture, Mathematics, and Health)* (Vol. 90). Springer.

Vande Zande, R. (2010). Teaching design education for cultural, pedagogical, and economic aims. *Studies in Art Education, 51*(3), 248–261.

Wang, H.-H., & Knobloch, N. A. (2018). Levels of STEM integration through agriculture, food, and natural resources. *Journal of Agricultural Education, 59*(3), 258–277.

Wannapiroon, N., & Petsangsri, S. (2020). Effects of STEAMification model in flipped classroom learning environment on creative thinking and creative innovation. *TEM Journal, 9*(4), 1647–1655.

Washuta, N. J., Eggleston, A. E., Righter, J., & Rabb, R. J. (2022). Defining key terms in new ABET student outcomes. In *2022 ASEE Annual Conference & Exposition*.

Yoh, T., Kim, J., Chung, S., & Chung, W. (2021). STREAM: A new paradigm for STEM education. *Journal of STEM Education: Innovations and Research, 22*(1), 46–51.

Zhao, Y. (2022). Systematic analysis of research trends in STEAM/STEM education based on big data. In *2022 International Conference on Educational Innovation and Multimedia Technology (EIMT 2022)* (pp. 155–168). Atlantis Press.

Zhou, C. (2012). Teaching engineering students creativity: A review of applied strategies. *Journal on Efficiency and Responsibility in Education and Science, 5*(2), 99–114.

Afterword

Confluence

Douglas Huffman, Imogen R. Herrick, and Kelli R. Feldman

The chapters in this book provide a compelling look at various ways the authors have reached out to students who are underrepresented in STEM. The ideas and projects described in this book have the promise to help transform STEM into a more equitable, inclusive, and welcoming place for all students. In the words of Professors Herrick and Lawson from chapter 1, there are "endless possibilities, but situational constraints." In other words, there are many intriguing ways to excite students about STEM and to draw them into the discipline; however, we should all be cautious about integrating other disciplines into STEM.

The STREAMS metaphor does not mean that one can simply add the arts, humanities and social sciences into STEM and assume that the integration will be effective. Instead, it offers ideas on what it means to create inclusive STEM opportunities and how that work can be approached. The power of these ideas does not lie within the acronyms or metaphors of STEM, STEAM, or STREAMS; instead, they lie within the ways teachers and students engage with them. However, engaging or integrating multiple disciplines can be quite challenging and requires carefully designed activities and curricula to effectively help students learn. We know that integrating the arts, humanities and social sciences can encourage students to engage in STEM, but we also know that teachers' ability to provide sufficient scaffolding for these integrations is not always effective or lasting.

The overall message from this book is that new approaches to encourage young students to engage in STEM can be thought-provoking and compelling; however, keeping students in STEM and encouraging them to pursue STEM in higher education and beyond is not straightforward. Many students enjoy STEM in elementary school but lose interest in STEM as they enter middle and high school. In addition, encouraging students who are typically underrepresented in STEM is even more challenging. Women and other underrepresented students do not find the fields of STEM as welcoming as other fields. The shortage of women and students from different ethnic and cultural backgrounds is concerning. STEM fields need to solve the underrepresentation problems for reasons of both workforce and building justice and sustainable futures.

Here, at the University of Kansas and in the state of Kansas, we recently celebrated the 70th anniversary of the *Brown v. Topeka Board of Education* historic supreme court decision. However, if you were to stand outside an elementary school in Topeka, Kansas, when the closing bell rings at the end of the day, you would not think that very much has changed regarding the racial and ethnic segregation of our schools. Today, as we did more than 70 years ago, we still have a disproportionate number of women and other underrepresented students in STEM. The numbers are slowly changing, but the lack of women and students of color in STEM is still a challenging issue to address. The issue is one of social justice. Underrepresented students in STEM have lower access, opportunity, and tend not to persist in the field. The persistent underrepresentation not only highlights systemic inequities but also underscores the importance of addressing the curricula's "discourse of politeness," which often glorifies STEM achievements while downplaying the detrimental impacts, such as industrial pollution of our air and water, and the toxic interactions that can occur through the use of high-tech social media (Rodriguez, 2015). By failing to critique the ethical implications of STEM applications, including environmental degradation and perpetuation of social inequalities, current standards like the NGSS and CCSSM avoid discussions that are crucial for the development of students' critical thinking and emotional intelligence (NGSS Lead States, 2013; Common Core State Standards, 2009). To truly diversify STEM fields and make them more inclusive, educational practices must integrate these critical discussions, challenging students to evaluate both the positive and negative impacts of scientific advancements and to envision STEM as a tool for social justice and ethical responsibility.

The future of STEM is full of possibilities, and it appears that the need for STEM professionals will continue to increase worldwide. With pressing challenges and potential related to artificial intelligence and climate change, it is apparent that we need more STEM professionals. We also need more STEM-literate citizens. Residents of planet earth need to have a basic understanding of STEM. As climate change, pollution, and technological advancements sweep the world, it becomes more and more important that all people have some basic understanding of STEM. As we read in the chapters of this book, there are "endless possibilities," but constraints such as time, resources, and the will of the people are disconcerting. We are hopeful that the winds of change will help to create new ways for all students to find purpose and joy in STEM. The STREAMS throughout the world provide a useful way to think about STEM and how we might open up STEM fields for all. We are at a nexus regarding climate change, and the solutions to problems our next generation will face require creativity, critical thinking, and hard work. We hope this STREAMS book will take us one step further toward envisioning a STEM world that is not only diverse, equitable, and inclusive but is also welcoming to all students in a way that helps them persevere into the future.

REFERENCES

Common Core State Standards. (2009). The core standards. Retrieved April 28, 2024, from https://www.thecorestandards.org

Next Generation Science Standards: For States, By States. (2013). The National Academies Press. https://doi.org/10.17226/18290

Rodriguez, A. J. (2015). What about dimensions of engagement, equity, and diversity practices? A critique of the next generation science standards. *Journal of Research in Science Teaching, 52*(7), 1031–1051.

About the Editors and Contributors

Douglas Huffman is a professor of science education at the University of Kansas. He has a B.S. in Civil Engineering from Stanford University, an M.Ed. in Education from Harvard University, and a Ph.D. in Science Education from the University of Minnesota. Dr. Huffman served as a co-editor of the *Journal of Research in Science Teaching* (JRST) and served as PI/Co-PI on numerous NSF grants, including the Collaborative Evaluation Communities in Urban Schools (CEC) and Linking Cognitive Science, Measurement, and Theory of Scientific Reasoning. Dr. Huffman's work is published in the *Journal of Research in Science Teaching, School Science and Mathematics, New Directions in Evaluation*, and the *Journal of Science Teacher Education*.

Kelli R. Feldman, Ph.D., is the Dean of the School of Education at Virginia Commonwealth University. Dr. Feldman co-edited a resource about transforming from STEM to STEAM education and served as the Co-PI on the NSF grant Collaborative Evaluation Communities in Urban Schools (CEC). She was also the mathematics content lead for the Department of Education grant Dynamic Learning Map Alternate Assessment System Consortium, and she served on the advisory board for Department of Education grants related to mathematics education. Dr. Feldman's work is published in the *Journal of Educational Research, School Science and Mathematics, Action in Teacher Education, Teaching Children Mathematics, American Journal of Evaluation, New Directions in Evaluation*, and *Teacher Education and Practice*.

Dr. Imogen R. Herrick is an assistant professor of STEM education at the University of Kansas. She focuses on two critical challenges in STEM education: how to integrate students' identities, communities, and experiences into their STEM learning and how to foster collaboration between students, teachers, and communities to address local environmental justice issues. Her approach involves exploring the intersection of emotion, place, and justice within STEM education, using participatory methods like photovoice, an effort that has earned her graduate student awards from the American Psychological Association (APA) and the American Educational Research Association. She is also a National Board Certified Science Teacher with 12 years of experience in K–12 settings. As an educator and researcher, she collaborates with teachers to improve and develop flexible instructional practices that focus on local justice-centered phenomena.

* * *

Clare Baek is a postdoctoral scholar at the Digital Learning Lab, University of California, Irvine. Her research focuses on promoting the learning and teaching of STEM skills for learners from diverse groups. Currently, she serves as a co-principal investigator of a National Science Foundation-funded project that aims to enhance the computational thinking of English learners through a computer science curriculum integrated with community-based environmental literacy and data literacy.

Amanda Bennett, Ph.D., is a lecturer in education and human development within the College of Education at Clemson University. Dr. Bennett serves as a faculty member in Clemson's undergraduate and master's level teacher education programs and teaches courses related to child development, educational psychology, classroom-based research, effective online teaching, and STEAM assessment strategies, to which she brings her own experience as a K–12 teacher. Her research interests focus on early childhood learning and development and STEAM (science, technology, engineering, arts/humanities, mathematics) instruction, specifically the integration and development of children's social-emotional skills, and teacher professional development. She has previously published a book focusing on the role of mobile devices and their impact on teacher professional development and has published articles in *The International Journal of Teacher Education and Professional Development*, *Elementary School Journal*, and *Mathematics Teacher: Learning and Teaching Pre-K–12*.

Britta Bletscher is a doctoral candidate in the University of Kansas's Curriculum and Teaching department. Britta's work focuses on studying the use of technology in first-year composition classes. She received her MA and BA in English from the University of Missouri-Kansas City. Britta is currently teaching full-time as an English instructor at Metropolitan Community

College-Longview and is a graduate teaching assistant for the University of Kansas.

Koeun Choi is an assistant professor of the Department of Human Development and Family Science at Virginia Tech. She directs the Cognitive Developmental Science (CoDeS) Lab. Her research examines the role of technologies, including touchscreens, artificial intelligence (AI) voice agents, and social robots, in children's attention and learning, with consideration of family processes and social contexts.

Prajnaparamita (Prajna) Dhar is a professor in the Department of Chemical and Petroleum Engineering at the University of Kansas. She also serves as a track director for the Bioengineering Graduate Program in the School of Engineering. Prajna's research interests include studying interfacial phenomena in biological and/or bioinspired systems and developing and implementing inclusive practices in undergraduate and graduate engineering education. Prajna collaborates with both scientists and engineers with diverse skills and experiences and proudly hosts a very diverse group of undergraduate and graduate students in her research lab. Her research work has been funded by both federal agencies (NSF and NIH) and industry partners (pharmaceutical companies). The products of her research are published in physics, chemistry, chemical engineering, or pharmaceutical chemistry focused journals, many of which are under the umbrella of the American Chemical Society or Elsevier Publishing Company.

Cara Eleonora Daza holds a Bachelor of Liberal Arts in Elementary Education from Bethel University and a Master of Education in Educational Leadership from High Tech High Graduate School of Education. Her diverse career spans public and private institutions in the United States and abroad. As an elementary STEM coordinator in Colombia, she coached teachers and students in PBL and STEM education, co-created an innovative robotics curriculum, and played a key role in achieving the first COGNIA STEM certification in South America. Additionally, she was recognized with a team award for the Digital Technology in Learning Award in 2021. Currently residing in Colombia, South America, Cara serves as the Program Director for Mentor a Mentor. In this role, she leads fundraising and leadership initiatives that support mentors and underserved youth throughout Latin America.

Sharin Jacob is a postdoctoral scholar in education at the University of California, Irvine. Her research examines the linguistic and sociocultural factors that help multilingual students succeed in computing. She has five years' experience teaching English as a second language, including launching a new arrivals center for newcomers. She was awarded the UC Irvine Public Impact Distinguished Fellowship for her commitment to bringing actionable change for multilingual students in computer science.

Myounghoon Jeon is an associate professor of the Department of Industrial and Systems Engineering and the Department of Computer Science (by courtesy) at Virginia Tech. His research focuses on human emotions and sound in the context of Human-Computer Interaction and Human-Robot Interaction. His research has yielded over 250 publications, including two books, *Emotions and Affect in Human Factors and Human-Computer Interaction* and *User Experience Design in the Era of Automated Driving*.

Leiny Yesenia Garcia is a PhD candidate at UC Irvine School of Education specializing in teaching, learning, and educational Improvement. Her work has focused on incorporating funds of knowledge and inclusive practices in STEM and CS education through the intersections of design-based implementation research and learning sciences.

Heidi L. Hallman is a professor and chair of the Department of Curriculum and Teaching at the University of Kansas. Hallman's research interests include how prospective teachers are prepared to teach in diverse school contexts as well as professional development opportunities for teachers. She is the co-author of *Secondary English Teacher Education in the United States* (2018), winner of the National Council of English's 2018 Richard A. Meade award for research in English education, among other works.

Michael Lawson is a teaching assistant professor of mathematics education at Kansas State University and vice president of the Kansas Teachers of Mathematics Association. Dr. Lawson's previous roles include high school mathematics teacher and secondary grades instructional coach in Knoxville, Tennessee, and postdoctoral fellow at the University of Southern California. His current research focuses on understanding and advancing in- and preservice teacher visioning for mathematics and STEM teaching and learning, along with developing and understanding small-scale, justice-centered routines that support teacher learning.

Santiago Ojeda-Ramirez is a Ph.D. student in the School of Education at the University of California, Irvine. He is interested in sociocultural approaches to learning for STEM and the arts. His work explores both tangible and digital construction for developing computational and AI literacies.

Meagan M. Patterson is a Professor of Educational Psychology at the University of Kansas. She has a B.A. in psychology and linguistics from the University of North Carolina at Chapel Hill and M.A. and Ph.D. degrees in developmental psychology from the University of Texas at Austin. Her research focuses on how children think about gender, race, and politics, and how adults talk with children about these topics, as well as inclusive teaching practices at the K-12 and higher education levels. Her work has been published in a variety of journals, including *Analyses of Social Issues and Public Policy, Journal of Youth and Adolescence, Monographs of the Society for Research in Child Development,* and *Social Psychology of Education.*

Gillian Roehrig is a professor of STEM education at the University of Minnesota. Her research explores issues of professional development for K–12 science teachers, with a focus on the implementation of integrated STEM learning environments and the induction and mentoring of beginning secondary science teachers. Her work in integrated STEM explores teachers' conceptions and implementation of STEM, curriculum development, and student learning in small groups during STEM lessons. She has received over $50 million in federal and state grants and published over 150 peer-reviewed journal articles and book chapters. She is a former president of the Association for Science Teacher Education and currently serves as the immediate past president of NARST.

Dana Saito-Stehberger is a research specialist and serves as the director of Curriculum and Professional Development on the Elementary Computing for all team at the University of California, Irvine. She has been in the English language teaching field for 30 years, teaching students and training teachers in intensive English programs, university courses, and in K–12 settings in the United States and internationally. Her research interests include the development of accessible and culturally relevant curricula for diverse learners as well as examining effective pedagogy for K–12 computer science education.

Mark Warschauer is a professor of education and director of the Digital Learning Lab at UC Irvine with affiliated appointments in informatics, language science, and psychological science. His research focuses on generative AI in education, the use of conversational agents to support children's learning, and the teaching and learning of computer science for linguistically diverse students. He is a member of the National Academy of Education.

www.ingramcontent.com/pod-product-compliance
Lightning Source LLC
Chambersburg PA
CBHW021144230426
43667CB00005B/253